2002 | Higher

[BLANK PAGE]

FOR OFFICIAL USE

Total for Sections B and C

X007/301

NATIONAL QUALIFICATIONS 2002

FRIDAY, 31 MAY
1.00 PM – 3.30 PM

BIOLOGY HIGHER

Fill in these boxes and read what is printed below.

Full name of centre

Town

Forename(s)

Surname

Date of birth
Day Month Year

Scottish candidate number

Number of seat

SECTION A—Questions 1–30 (30 marks)

Instructions for completion of Section A are given on page two.

SECTIONS B AND C (100 marks)

1. (a) All questions should be attempted.
   (b) It should be noted that in **Section C** questions 1 and 2 each contain a choice.
   (c) Question 9 is on pages 22, 23 and 24. Question 10 is on pages 25, 26 and 27. Pages 24 and 25 are fold-out pages. The additional graph paper is on page 36.
2. The questions may be answered in any order but all answers are to be written in the spaces provided in this answer book, and must be written clearly and legibly in ink.
3. Additional space for answers and rough work will be found at the end of the book. If further space is required, supplementary sheets may be obtained from the invigilator and should be inserted inside the **front** cover of this book.
4. The numbers of questions must be clearly inserted with any answers written in the additional space.
5. Rough work, if any should be necessary, should be written in this book and then scored through when the fair copy has been written.
6. Before leaving the examination room you must give this book to the invigilator. If you do not, you may lose all the marks for this paper.

SCOTTISH QUALIFICATIONS AUTHORITY

SECTION A

Read carefully

1. Check that the answer sheet provided is for Biology Higher (Section A).
2. Fill in the details required on the answer sheet.
3. In this section a question is answered by indicating the choice A, B, C or D by a stroke made in **ink** in the appropriate place in the answer sheet—see the sample question below.
4. For each question there is only **one** correct answer.
5. Rough working, if required, should be done only on this question paper—or on the rough working sheet provided—**not** on the answer sheet.
6. At the end of the examination the answer sheet for Section A **must** be placed inside the front cover of this answer book.

Sample Question

The apparatus used to determine the energy stored in a foodstuff is a

A respirometer

B calorimeter

C klinostat

D gas burette.

The correct answer is **B**—calorimeter. A **heavy** vertical line should be drawn joining the two dots in the appropriate box in the column headed **B** as shown in the example on the answer sheet.

If, after you have recorded your answer, you decide that you have made an error and wish to make a change, you should cancel the original answer and put a vertical stroke in the box you now consider to be correct. Thus, if you want to change an answer D to an answer B, your answer sheet would look like this:

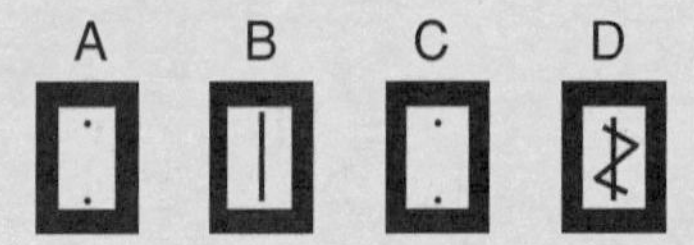

If you want to change back to an answer which has already been scored out, you should enter a tick (✓) to the **right** of the box of your choice, thus:

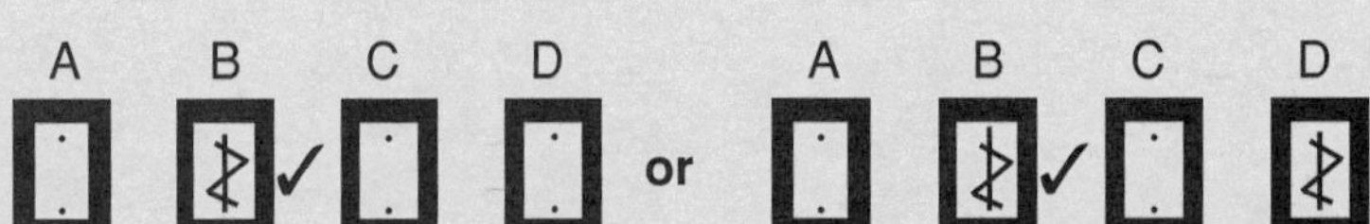

SECTION A

All questions in this section should be attempted.

Answers should be given on the separate answer sheet provided.

1. When an animal cell is immersed in a hypotonic solution it will

A burst

B become turgid

C shrink

D become flaccid.

2. In photosynthesis, the function of the pigment carotene is to

A receive light energy from chlorophyll for photolysis of water

B allow the plant to absorb a wider range of wavelengths of light

C allow photosynthesis to take place in light of low intensity

D increase the capacity of chlorophyll to absorb light.

3. On Earth, the energy input from sunlight is around $2 \cdot 0 \times 10^{12}$ kilojoules per hectare per annum. The energy captured by photosynthesising plants is around $2 \cdot 0 \times 10^{10}$ kilojoules per hectare per annum. (A hectare is a measurement of area.)

What is the percentage efficiency of photosynthesis of these plants?

A 1%

B 2%

C 17%

D 200%

4. Three different strains of yeast each lacked a different respiratory enzyme involved in the complete breakdown of glucose.

Strain X – cannot produce carbon dioxide from pyruvic acid.

Strain Y – cannot form pyruvic acid.

Strain Z – cannot reduce oxygen to form water.

Which of the strains could produce ethanol?

A Strains X and Y

B Strains X and Z

C Strain Y only

D Strain Z only

5. Which of the following proteins has a fibrous structure?

A Pepsin

B Amylase

C Insulin

D Collagen

6. Insulin synthesised in a pancreatic cell is secreted from the cell. Its route from synthesis to secretion includes

A Golgi apparatus → endoplasmic reticulum → ribosome

B ribosome → Golgi apparatus → endoplasmic reticulum

C endoplasmic reticulum → ribosome → Golgi apparatus

D ribosome → endoplasmic reticulum → Golgi apparatus.

7. If a living tissue is transplanted from one person to another, there is a risk of rejection because the recipient reacts against the foreign

A antibodies

B antigens

C DNA

D RNA.

[Turn over

8. Which of the following cells are responsible for producing antibodies?

A Monocytes

B Lymphocytes

C Phagocytes

D Red blood cells

9. Viruses consist of a

A lipid coat enclosing DNA or RNA

B protein coat enclosing DNA only

C protein coat enclosing DNA or RNA

D lipid coat enclosing DNA only.

10. What is the significance of chiasma formation?

A It results in the halving of the chromosome number.

B It results in the pairing of homologous chromosomes.

C It permits gene exchange between homologous chromosomes.

D It results in the independent assortment of chromosomes.

11. The table below shows some genotypes and phenotypes associated with forms of sickle-cell anaemia.

Genotype	*Phenotype*
$Hb^A\ Hb^A$	normal
$Hb^A\ Hb^S$	sickle-cell trait
$Hb^S\ Hb^S$	acute sickle-cell anaemia

A normal man marries a woman with the sickle-cell trait. What are the chances that any child born to them will have acute sickle-cell anaemia?

A None

B 1 in 1

C 1 in 2

D 1 in 4

12. The diagram shows the transmission of the gene for albinism.

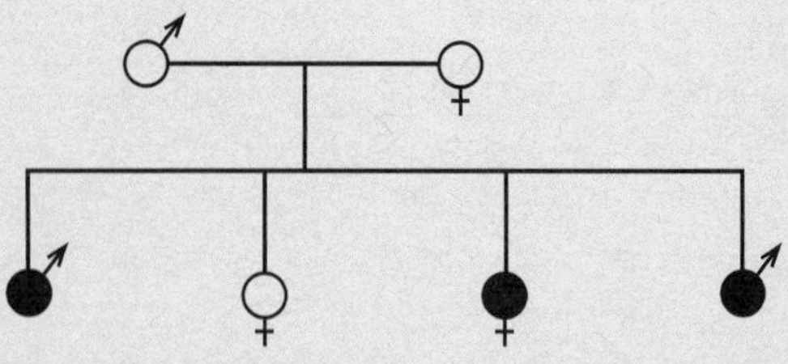

Key

normal male ♂ affected male

normal female ♀ affected female

This condition is inherited as a characteristic which is

A dominant and not sex-linked

B recessive and not sex-linked

C dominant and sex-linked

D recessive and sex-linked.

13. In *Drosophila*, the long-winged condition (L) is dominant to the vestigial-winged condition (l) and broad abdomen (B) is dominant to narrow abdomen (b).

When parent flies, heterozygous for both wing shape and abdomen width, were crossed with flies having vestigial wings and narrow abdomens, the results were as shown in the table below.

	Number of male offspring	*Number of female offspring*
Long wings, broad abdomen	230	227
Long wings, narrow abdomen	4	3
Vestigial wings, broad abdomen	3	3
Vestigial wings, narrow abdomen	238	240

These results indicate

A crossing over

B independent assortment

C a mutation

D sex-linkage.

14. Colour blindness is a recessive, sex-linked characteristic controlled by the allele b.

Two parents with normal vision have a colour-blind boy.

The genotypes of the parents are

A X^BY and X^BX^b

B X^bY and X^BX^B

C X^bY and X^BX^b

D X^BY and X^bX^b.

15. Which of the following mutations would cause a change in chromosome number?

A Translocation

B Non-disjunction

C Inversion

D Insertion

16. Genes **a** to **j** occur on part of a chromosome.

a	b	c	d	e	f	g	h	i	j

After cell division, this part of the chromosome had the following sequence of genes:

a	b	c	d	e	f	g	d	e	f	g	h	i	j

This change is called a

A repetition

B translocation

C duplication

D replication.

17. Cabbage and radish each have a diploid number of 18. These plants can be crossed to produce a hybrid.

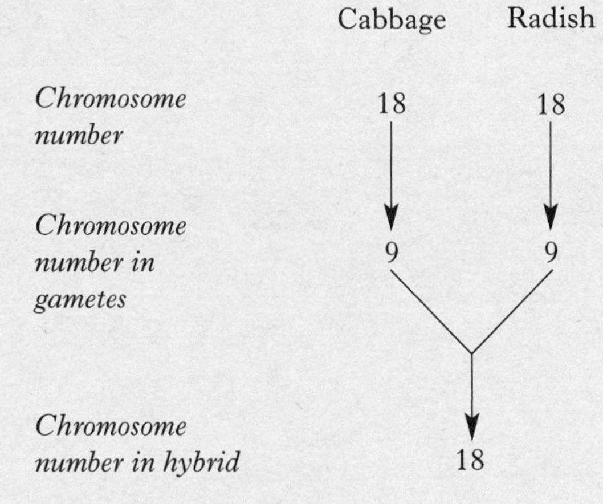

Which of the following statements is true?

A The hybrid contains 9 pairs of homologous chromosomes.

B The chromosomes in the hybrid cannot pair during meiosis.

C The hybrid can produce gametes with a haploid number of 9.

D The hybrid can interbreed successfully with either parent.

18. In the formation of protoplasts, plant cells are treated with

A amylase

B lipase

C restriction enzymes

D cellulase.

19. Which of the following is an example of intraspecific competition?

A Plants of the same species competing for the same growth requirements.

B Plants of the same species competing for different growth requirements.

C Plants of a different species competing for the same growth requirements.

D Plants of a different species competing for different growth requirements.

[Turn over

20. In an animal, habituation has taken place when a

A harmful stimulus ceases to produce a response

B harmful stimulus always produces an identical response

C harmless stimulus always produces an identical response

D harmless stimulus ceases to produce a response.

21. The rates of carbon dioxide exchange by the leaves of two species of plants were measured at different light intensities.

The results are shown in the graph below.

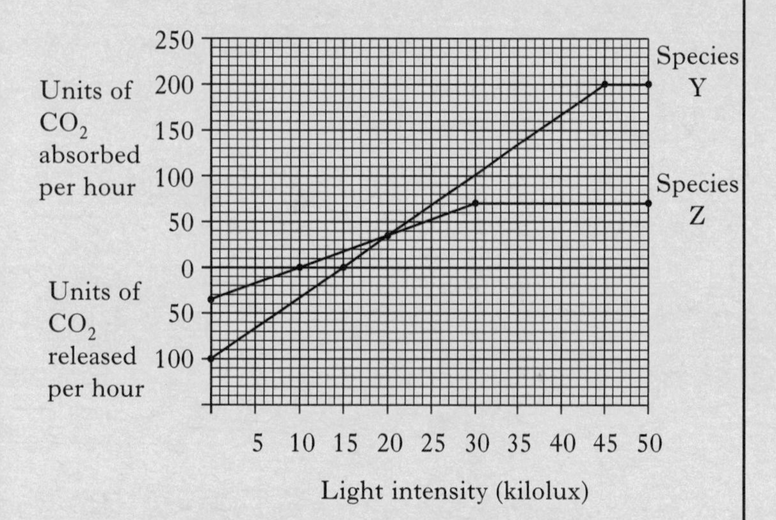

What are the light intensities at which species Z and Y reach their compensation points?

	Light Intensity (kilolux)	
	Z	Y
A	10	15
B	20	20
C	20	30
D	30	45

22. Which of the following statements explains the structural differences between cells in different tissues of an organism?

A Cells in some tissues have more genes than cells in other tissues.

B As different tissues develop, different genes are lost from their cells.

C Different cell types have the same genes but different genes are active.

D Some tissues have genes from one parent while some have genes from the other.

23. The thyroid gland is involved in the control of metabolic rate.

Which of the following shows the correct sequence for metabolic control.

A Pituitary → thyroxine → thyroid → TSH

B Pituitary → TSH → thyroid → thyroxine

C TSH → thyroxine → pituitary → thyroid

D Thyroid → TSH → thyroxine → pituitary

24. Which of the following is the correct sequence of events that occurs in control of the concentration of blood sugar?

	Concentration of blood sugar	*Glucagon secretion*	*Insulin secretion*	*Glycogen stored in liver*
A	increases	decreases	increases	increases
B	increases	decreases	increases	decreases
C	decreases	increases	decreases	increases
D	decreases	decreases	increases	decreases

25. Which of the following is **not** the result of a magnesium deficiency in flowering plants?

A Curling of the leaves

B Yellowing of the leaves

C Reduction in shoot growth

D Reduction in root growth

26. A 30 g serving of breakfast cereal with 125 cm^3 of semi-skimmed milk contains 1·5 mg of iron. Only 25% of this iron is absorbed into the bloodstream.

If a woman in late pregnancy requires a daily intake of 6 mg of iron, how much cereal and milk would have to be eaten to meet this requirement?

	Cereal (g)	*Milk* (cm^3)
A	60	250
B	120	500
C	240	1000
D	480	2000

27. In humans, vitamin D plays an essential role in the absorption of

A glucose

B calcium

C iron

D lipids.

28. Which one of the following factors affecting a population of rabbits is density independent?

A Viral disease

B The population of foxes

C The biomass of the grass

D Rainfall

29. The diagram shows the average lifespan of people in Britain between 1900 and 1990.

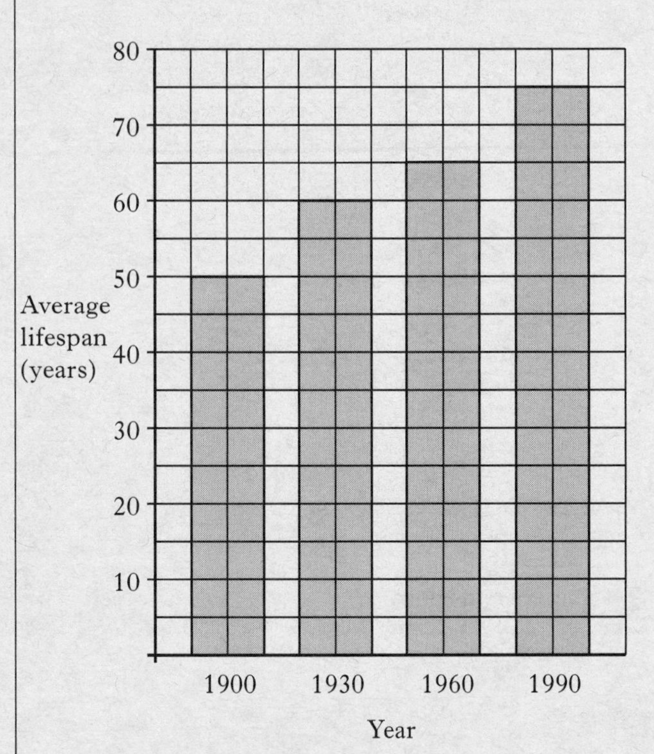

What is the percentage increase in lifespan during this period?

A 25%

B 45%

C 50%

D 75%

30. The graphs below contain information about the population of Britain.

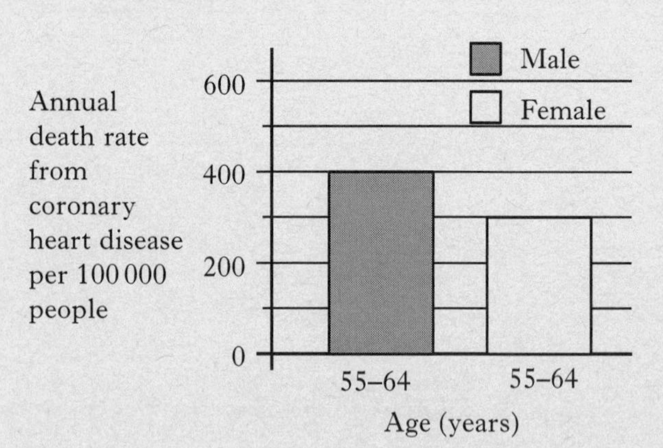

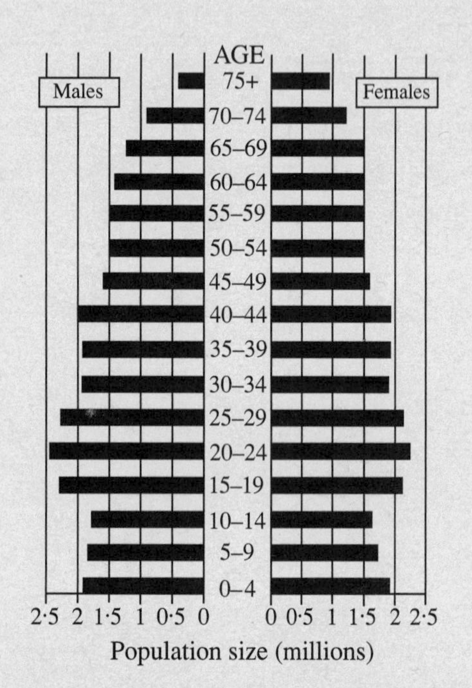

How many British women between 55 and 64 years of age die from coronary heart disease annually?

A 300

B 4500

C 9000

D 21 000

Candidates are reminded that the answer sheet MUST be returned INSIDE the front cover of this answer book.

[Turn over for Section B on *Page ten*

DO NOT WRITE IN THIS MARGIN

Marks

SECTION B

All questions in this section should be attempted.

1. The diagram below represents stages in aerobic respiration in mammalian liver cells.

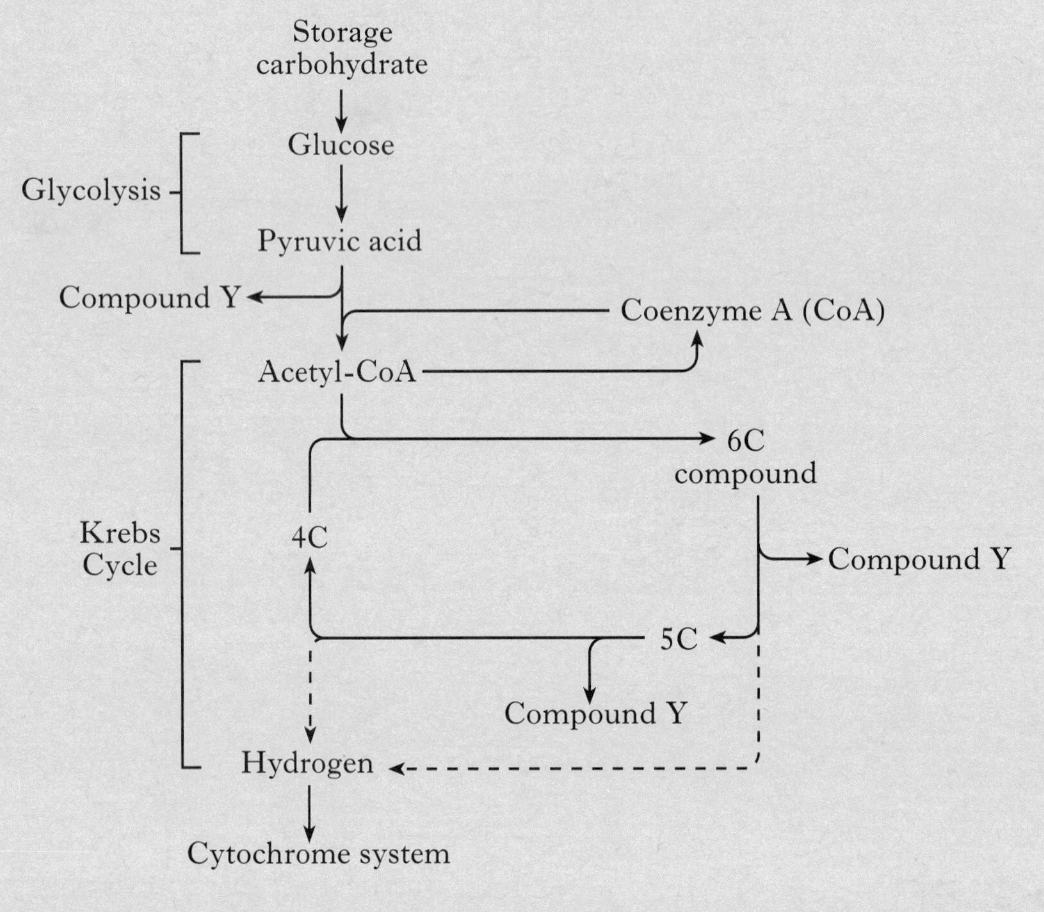

(*a*) Name the storage carbohydrate in liver cells.

________________________ **1**

(*b*) Other than carbohydrate, name an alternative respiratory substrate.

____ **1**

(*c*) State the exact location within a liver cell of:

1 Glycolysis ________________________

2 Krebs cycle ________________________ **2**

(*d*) State the net gain of ATP molecules from the breakdown of a glucose molecule to pyruvic acid.

________________________ **1**

DO NOT WRITE IN THIS MARGIN

1. (continued) *Marks*

(*e*) Name the 6 carbon compound and compound Y shown in the Krebs cycle.

6C compound ____________________

Compound Y ____________________ **2**

(*f*) Name the chemical that transports hydrogen from the Krebs cycle to the cytochrome system.

____ **1**

[Turn over

DO NOT WRITE IN THIS MARGIN

Marks

2. (*a*) The graph below shows the effects of varying light intensity, carbon dioxide concentration and temperature on the rate of photosynthesis in a plant.

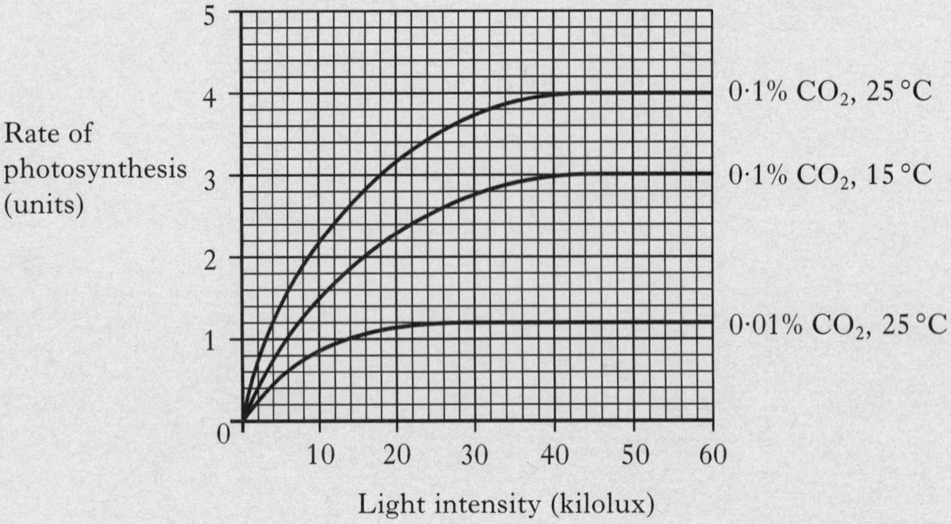

Rate of photosynthesis (units)

0·1% CO_2, 25 °C

0·1% CO_2, 15 °C

0·01% CO_2, 25 °C

Light intensity (kilolux)

(i) At a light intensity of 60 kilolux, identify which factor, as shown in the graph, has the greater effect in increasing the rate of photosynthesis. Justify your answer.

Factor ______________________

Justification ______________________

____

______________________ **1**

(ii) The experiment was repeated at a carbon dioxide concentration of 0·01% and a temperature of 15 °C. **Onto the graph above**, draw a curve to show the predicted results of this experiment. **1**

(iii) The rate of photosynthesis can be calculated by measuring a change in dry mass with time.
State **one** other method that can be used to calculate the rate of photosynthesis.

____

______________________ **1**

(*b*) Other than being absorbed for use in photosynthesis, give **two** possible fates of the light energy that shines onto the leaves of plants.

1 ______________________

2 ______________________ **1**

(*c*) Within which region of a chloroplast does the absorption of light take place?

______________________ **1**

DO NOT WRITE IN THIS MARGIN

2. (continued)

Marks

(*d*) Name the **two** chemical compounds produced in the light dependent stage that are essential for the conversion of GP to glucose.

Compound 1 ______________________

Compound 2 ______________________ **1**

(*e*) Complete the table below by inserting the number of carbon atoms present in each of the chemical compounds.

Compound	*Number of carbon atoms*
GP	
RuBP	
Glucose	

1

(*f*) What would be the effect on the RuBP concentration if conditions were changed from light with carbon dioxide present to light with carbon dioxide absent.
Explain your answer.

Effect ______________________

Explanation ______________________

____

______________________ **2**

(*g*) The list below shows terms which refer to the role of light in the growth and development of plants and animals.

List of terms
Phototropism
Compensation point
Etiolation
Photoperiodism

In the table below insert the term that is described by each statement.

Statement	*Term*
The rate of respiration equals the rate of photosynthesis at a certain light intensity.	
If germinated in darkness, a seedling has long internodes and small yellow leaves.	
Plants respond to directional light by a growth curvature.	
A change in the number of hours of light in the day can affect growth and development in many plants and animals.	

2

DO NOT WRITE IN THIS MARGIN

Marks

3. (*a*) Bar graph 1 shows the concentrations of sodium and potassium in the cell sap of a water plant and in the surrounding pond water.
The data were collected at 20 °C under aerobic conditions.

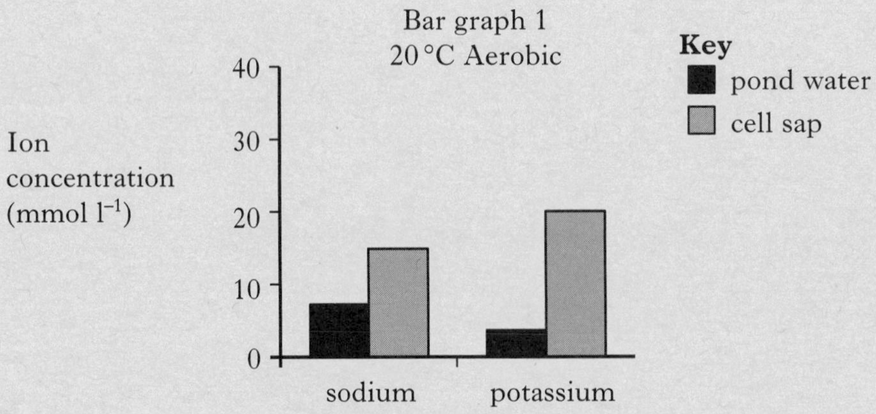
Bar graph 1
20 °C Aerobic
Key
pond water
cell sap
Ion concentration (mmol l⁻¹)
40
30
20
10
0
sodium
potassium

(i) What evidence from the bar graph suggests that:

1 ion uptake is by active transport? ______________________

____ **1**

2 ion uptake is selective? ______________________

____ **1**

The experiment was repeated at different temperatures and oxygen availability. The results obtained are shown below.

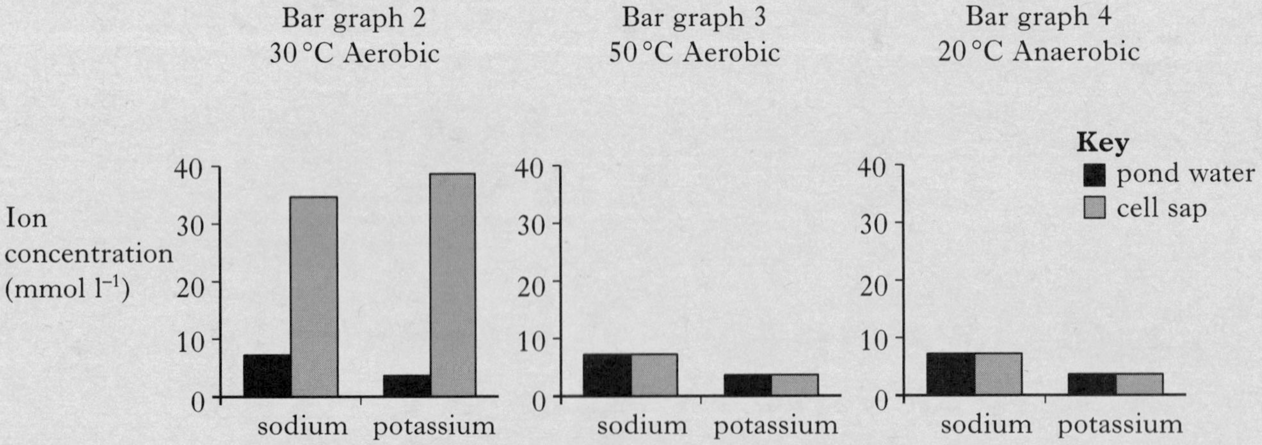
Bar graph 2
30 °C Aerobic
Bar graph 3
50 °C Aerobic
Bar graph 4
20 °C Anaerobic
Key
pond water
cell sap
Ion concentration (mmol l⁻¹)
40
30
20
10
0
sodium
potassium

DO NOT WRITE IN THIS MARGIN

Marks

3. (*a*) (continued)

(ii) Complete the table below to identify **two** bar graphs which could be compared to support each statement.

Statement	*Comparison*	
	Bar graph number	Bar graph number
Enzyme activity increases with increasing temperature		
Membrane proteins may be denatured		

2

(iii) Account for the differences in ion uptake shown in bar graphs 1 and 4.

____

____

______________________________ **2**

(*b*) State the importance of potassium in the normal functioning of plant cells.

____

______________________________ **1**

[Turn over

DO NOT WRITE IN THIS MARGIN

Marks

4. (*a*) A gamete mother cell undergoes meiosis to form gametes.
The bar graph below shows the DNA content per cell at different stages in meiosis.

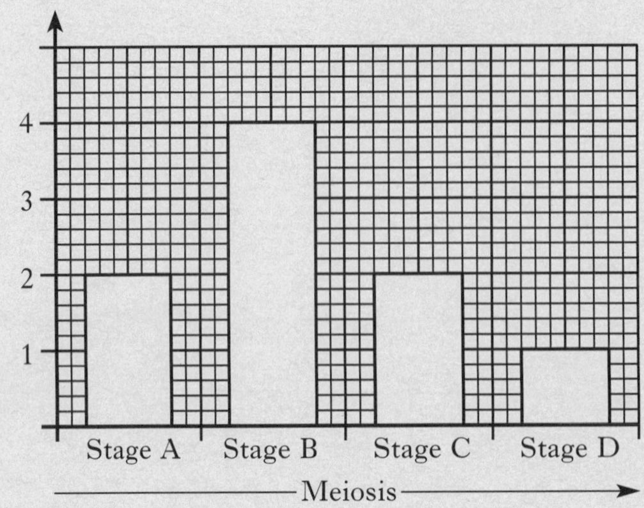

Describe what happens during meiosis to account for the change in the DNA content per cell between the following stages.

1 Stages A and B ______________________________ **1**

2 Stages B and C ______________________________ **1**

3 Stages C and D ______________________________ **1**

(*b*) The table below shows the percentage recombination frequencies for four genes present on the same chromosome.

Gene pairs	*% recombination frequency*
P and Q	16
P and R	8
R and S	12
Q and S	4

(i) What term is used to describe genes present on the same chromosome?

______________________________ **1**

(ii) Use the information to identify the order in which these genes lie on the chromosome.

Space for calculation

Order ______________________________ **1**

DO NOT WRITE IN THIS MARGIN

Marks

5. (*a*) Cyanogenic clover plants produce cyanide when their tissues are damaged. Cyanide is toxic and its production defends plants against herbivores.
The diagram below shows the metabolic pathway which produces cyanide.

Substrate A ---→ Substrate B ---→ cyanide
↑ Enzyme 1 ↑ Enzyme 2

The genes which code for Enzymes 1 and 2 have alleles with properties shown in the table below.

Gene coding for Enzyme 1		*Gene coding for Enzyme 2*	
Allele	Synthesises Enzyme 1	Allele T	Synthesises Enzyme 2
Allele r	Cannot synthesise Enzyme 1	Allele t	Cannot synthesise Enzyme 2

(i) Complete the cross below by:

1 inserting the genotype of the gametes of the individual with the genotype RrTt;

2 showing the possible genotypes of the offspring.

Parental phenotypes	Cyanogenic	X	Non-cyanogenic
Parental genotypes	RrTt	X	rrtt

1

1

(ii) Express the expected number of cyanogenic to non-cyanogenic plants as the simplest whole number ratio.

Ratio cyanogenic _______ : non-cyanogenic _______ 1

(*b*) Give **two** examples of adaptions which allow plants to tolerate grazing.

Example 1 ________________________________

Example 2 ________________________________ 1

[Turn over

DO NOT WRITE IN THIS MARGIN

Marks

6. (*a*) The following statements refer to the Hawaiian islands and species which inhabit them.

Statements

1. The Hawaiian islands are of volcanic origin.
2. The islands are far from any continental mainland.
3. 91% of the plant species and 81% of the bird species are found only on these islands.
4. Hawaiian honeycreepers are species of birds which are descended from a seed-eating ancestral species.
5. Honeycreeper species show a wide range of beak shapes for eating the seeds, fruits, nectar and insects available.
6. Estimates of the present number of honeycreeper species range from 29 to 33 with many extinctions having occurred after the arrival of man on the islands.

(i) Name the isolation mechanism illustrated in Statement 2.

____ **1**

(ii) State the importance of isolating mechanisms in the evolution of new species.

____

____

____ **1**

(iii) Identify the **two** statements which suggest that the evolution of the honeycreeper species is an example of adaptive radiation.

Statement numbers __________ and __________ **1**

(iv) Name **two** methods used to conserve species and prevent their extinction. (Statement 6)

____

____

____ **2**

DO NOT WRITE IN THIS MARGIN

Marks

6. (continued)

(*b*) Populations may be monitored to provide data for a wide variety of purposes.

The table below shows the sulphur dioxide concentrations within various areas and the number of lichen species present.

Sulphur dioxide concentration in area (ppm)	*Number of lichen species present*
32	0
17	4
8	11
0	17

From the table explain how lichen can be used as indicator species.

1

(*c*) Explain the need to monitor populations of fish such as cod.

1

[Turn over

DO NOT WRITE IN THIS MARGIN

Marks

7. The bacterium *Bacillus thuringiensis* produces a substance called T-toxin that is harmful to leaf-eating insects.

The information below shows some of the procedures used by genetic engineers to insert the gene for the production of T-toxin into crop plants.

Procedure 1 Chromosome extracted from bacterial cells

↓

Procedure 2 Position of T-toxin gene located

↓

Procedure 3 T-toxin gene cut out from bacterial chromosome

↓

Procedure 4 T-toxin gene transferred into nucleus of host plant cell

↓

Procedure 5 Plant cells containing T-toxin gene grown into plantlets

(*a*) Name a technique that could be used in Procedure 2 to locate the position of the T-toxin gene.

____ **1**

(*b*) Name the enzyme used in procedure 3.

____ **1**

(*c*) Explain why such genetically engineered crop plants would grow better than unmodified crop plants.

____

____

____ **2**

(*d*) These crops were commercially successful for several years. However, they have since become susceptible to attack by some members of a particular insect species.
Suggest a reason that would account for this observation.

____

____ **1**

DO NOT WRITE IN THIS MARGIN

Marks

8. The flowchart represents the control system involved in returning body temperature to normal after an increase.

Increase in body temperature
↓
Increase detected by temperature-monitoring centre
↓
Corrective mechanisms switched on
↓
Return to normal body temperature
↓
Decrease detected by temperature-monitoring centre
↓
Corrective mechanisms switched off

(*a*) State the exact location of the temperature-monitoring centre.

____ **1**

(*b*) In the following sentence **underline** one of the alternatives in each pair to describe a corrective mechanism that is switched on when body temperature increases.

In this corrective mechanism {vasoconstriction / vasodilation} results in {increased / decreased} blood flow to the skin and therefore, {increased / decreased} heat loss from the skin by radiation. **1**

(*c*) Describe another corrective mechanism that would reduce body temperature.

____

____ **1**

(*d*) How is the message carried from the temperature-monitoring centre to effectors?

____ **1**

(*e*) What is the importance of body temperature in humans to metabolic processes?

____

____ **1**

DO NOT WRITE IN THIS MARGIN

9. An investigation was carried out into the effect of different concentrations of indole-acetic acid (IAA) on growth of shoot tips.

The method used in the investigation is outlined below.

1. 10 mm lengths of shoot tissue were cut from behind the shoot tip meristem.
2. Five lengths of shoot tissue were used in each experiment.
3. Shoot tissues were immersed in solutions containing different concentrations of IAA.
4. The diagram below shows the experimental set up at one of the concentrations of IAA.

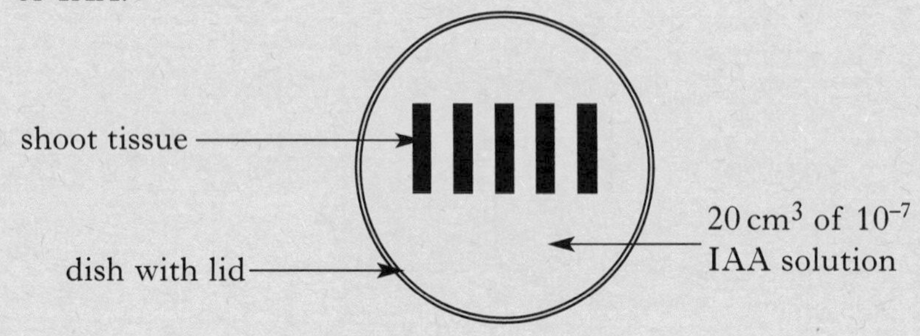

5. A control experiment was set up with five 10 mm lengths of intact shoot tip tissue immersed in distilled water.
6. The experiments were left in the dark for 48 hours.
7. The length of each shoot tip tissue was measured.
8. For each IAA concentration, the average length of shoot tip tissue was compared with the control experiment.

Key

+ Growth greater than control

– Growth less than control

The results are shown in the table below.

Concentration of IAA solution (molar)	*Difference between the average length of shoot tissue and the control* (mm)
10^{-7}	+4
10^{-6}	+8
10^{-5}	+5
10^{-4}	+3
10^{-3}	0
10^{-2}	–2
10^{-1}	–4

[Question 9 continues on *Page twenty-three* and fold-out *Page twenty-four*

DO NOT WRITE IN THIS MARGIN

Marks

9. (continued)

(*a*) On the grid below, using a suitable scale, plot a line graph of difference between the average length of shoot tissue and the control against concentration of IAA solution.
(Additional graph paper, if required, can be found on *Page thirty-two*.)

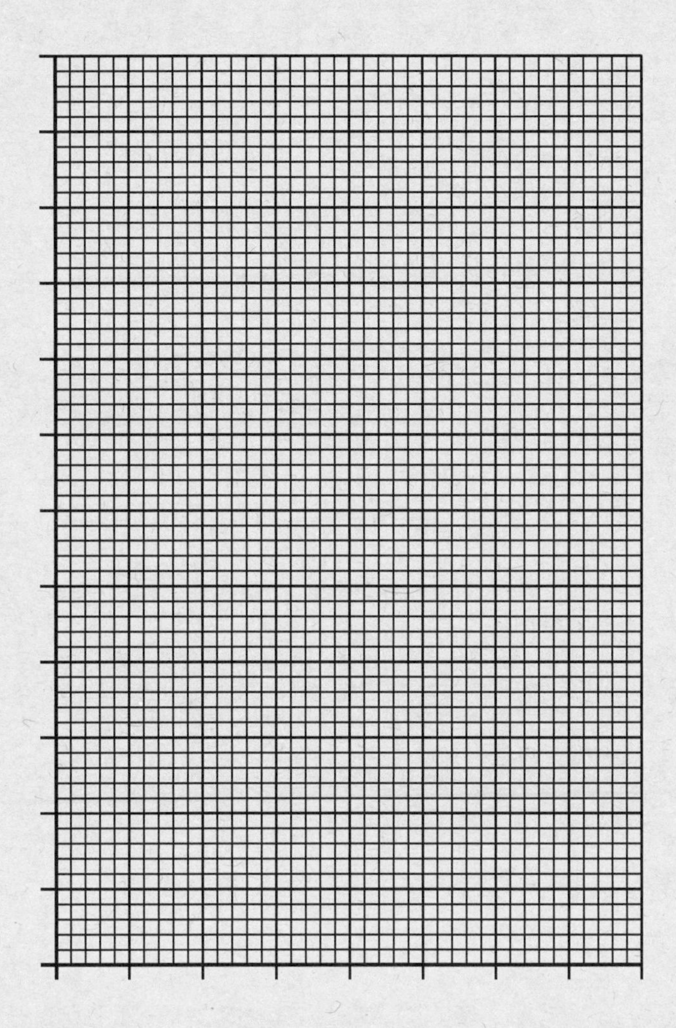

2

(*b*) Describe the pattern of growth of shoot tissue over the range of IAA concentrations used in the investigation.

____

____

____ **2**

(*c*) Why is it considered good experimental procedure to leave the shoot tissues in the solutions for 48 hours?

____

____ **1**

[Question 9 continues on *Page twenty-four*

DO NOT WRITE IN THIS MARGIN

Marks

9. (continued)

(*d*) All the IAA solutions were measured using the same syringe. What precaution should be taken to minimise experimental error?

____

______________________________ **1**

(*e*) In this investigation what would appear to be the naturally occurring concentration of IAA within the shoot tip?

____

______________________________ **1**

[Question 10 begins on fold-out *Page twenty-five*

DO NOT WRITE IN THIS MARGIN

10. The red spider mite is a pest of crop plants. Stages in its life-cycle are shown in the diagram below.

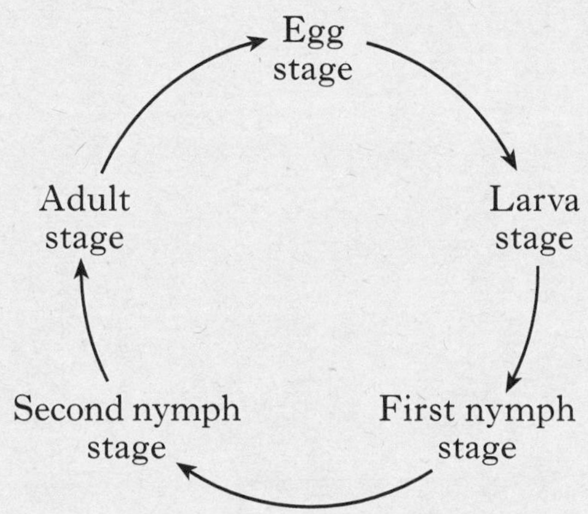

An investigation was carried out into the effects of temperature on the time each of the early stages lasts. The results are shown in the bar graph below.

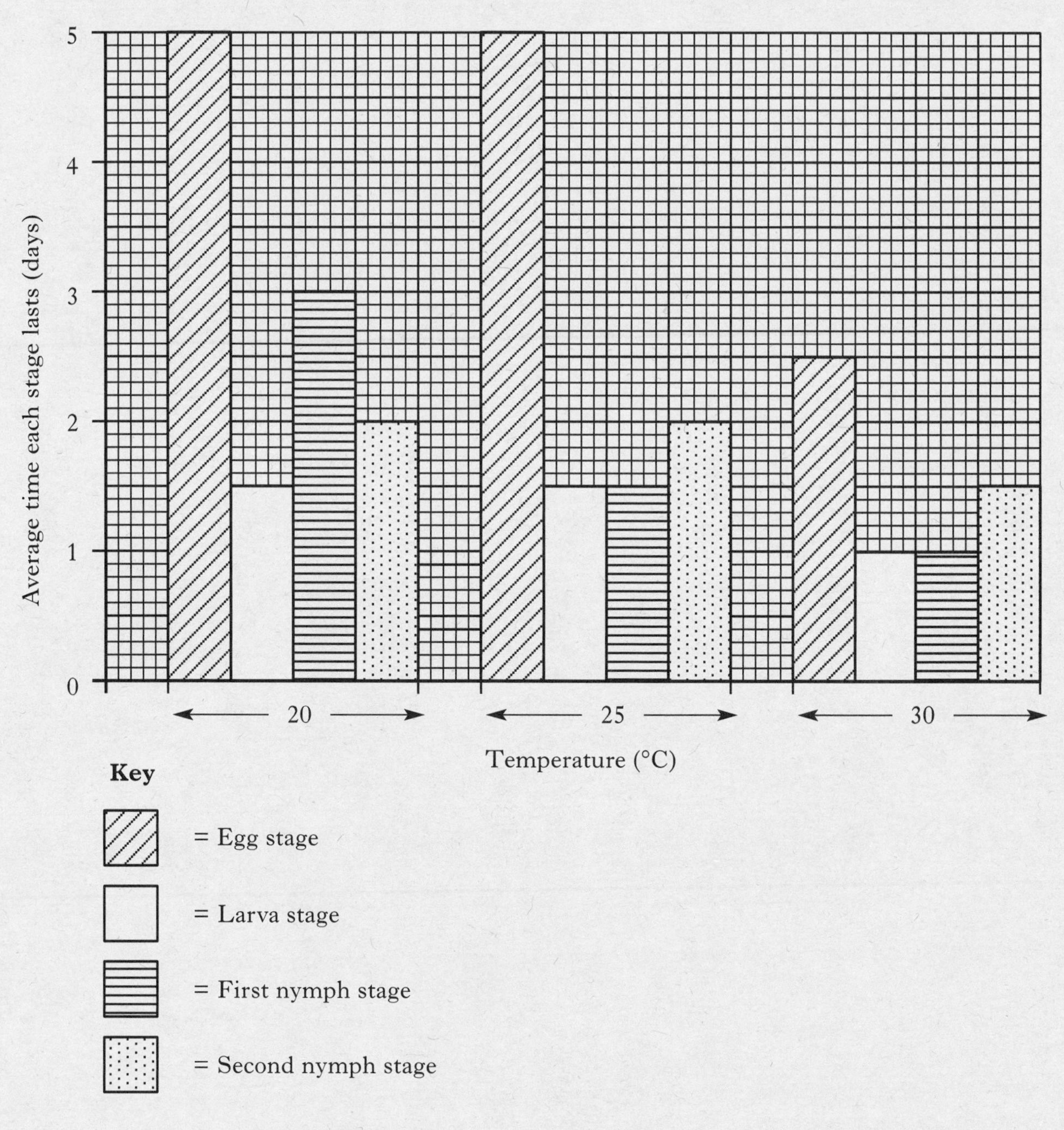

Marks

10. (continued)

(*a*) From the bar graph, calculate the difference in average time for development through the egg stage to the start of the adult stage at 20 °C and 30 °C.

Space for calculation

Difference ________ days **1**

Table 1 below shows the effect of temperature on features of egg laying in adult red spider mites.

Table 1

Features	*Temperature (°C)*		
	20 °C	*25 °C*	*30 °C*
Average time spent as an adult before first egg laid (days)	2·20	1·60	1·00
Average length of egg laying period (days)	18·40	14·80	10·36
Average number of eggs laid per female during egg laying period	92·00	88·80	62·16

(*b*) From Table 1, describe the relationship between temperature and each feature of egg-laying in adult female red spider mites.

____

____

____

____

______________________________ **2**

(*c*) From Table 1, calculate the average number of eggs laid per female per day during the egg laying period at 20 °C.

Space for calculation

Average number of eggs laid per female per day ________ **1**

DO NOT WRITE IN THIS MARGIN

Marks

10. (continued)

(*d*) From Table 1, calculate the percentage decrease in the average number of eggs laid per female during the egg laying period when the temperature is increased from 25 °C to 30 °C.

Space for calculation

% decrease__________ **1**

(*e*) From the information in the bar graph and Table 1, complete the table below by writing **True** or **False** in each of the spaces provided.

Statement	*True/False*
The time for the length of the first nymph stage is shortest at 30 °C.	
The egg stage lasts twice as long at 30 °C as at 25 °C.	
Only the first nymph stage is affected by a change from 20 °C to 25 °C.	
At 20 °C some adults may take more than 2·2 days to start laying eggs.	

2

(*f*) From the bar graph and Table 1, calculate the average time that it takes for development from the egg stage to the laying of the first egg at 25 °C.

Space for calculation

Average time __________ days **1**

[Turn over

Marks

11. (*a*) The graph below shows changes in stomatal width over a 24-hour period of a plant species that is adapted to live in a hot climate.

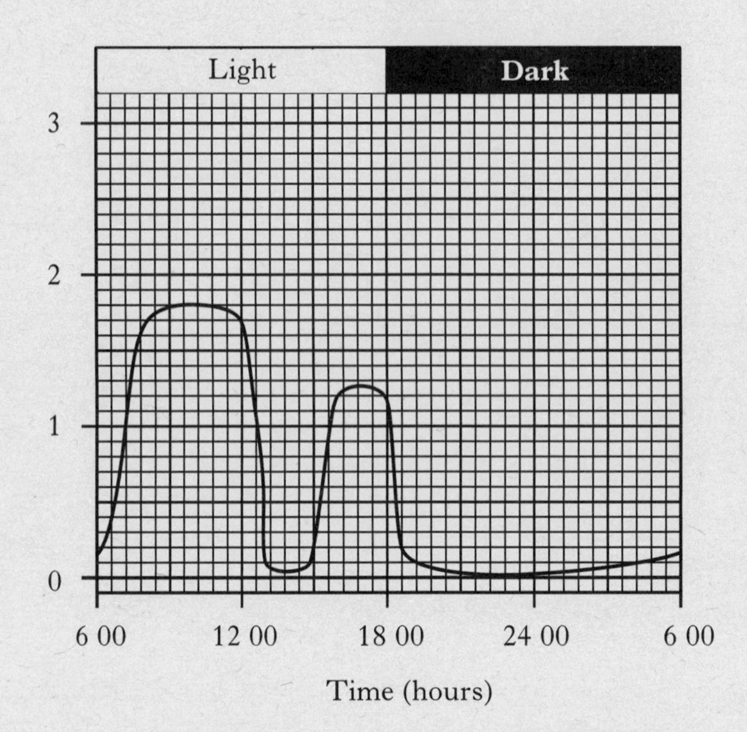

(i) State the change in turgor that takes place in the guard cells to cause stomatal closure.

____ **1**

(ii) 1 Explain how the pattern of change in stomatal width between 11 00 and 16 00 may benefit a plant that lives in a hot climate.

____

____

____ **2**

2 Suggest a possible disadvantage to the plant of this pattern of change in stomatal width.

____

____ **1**

DO NOT WRITE IN THIS MARGIN

11. (continued) *Marks*

(*b*) Plants that live in desert conditions have adaptations which reduce water loss. The diagram below shows part of a leaf section of a desert plant.

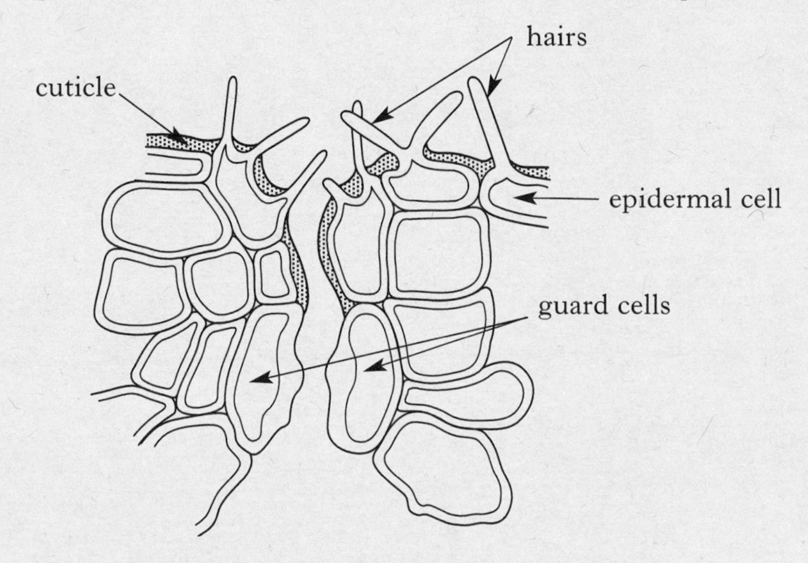

Complete the table below which shows leaf adaptations and explanations of how these reduce water loss.

Description of adaptation	*Explanation for reduction in water loss*
Presence of hairs on the leaf surface	
	Longer distance for water vapour to diffuse out of the leaf.

2

(*c*) Salmon and eels have adaptations associated with migration between freshwater and seawater.

(i) State the change that takes place in the glomerular filtration rate of these fish when they return to the sea.

______________________________ 1

(ii) Describe the role of the chloride secretory cells when these fish are in seawater.

______________________________ 1

(*d*) Describe **one** behavioural and **one** physiological adaptation shown by the desert rat to reduce water loss.

Behavioural ______________________________

______________________________ 1

Physiological ______________________________

______________________________ 1

DO NOT WRITE IN THIS MARGIN

Marks

12. (*a*) The diagram below represents a section through a woody stem.

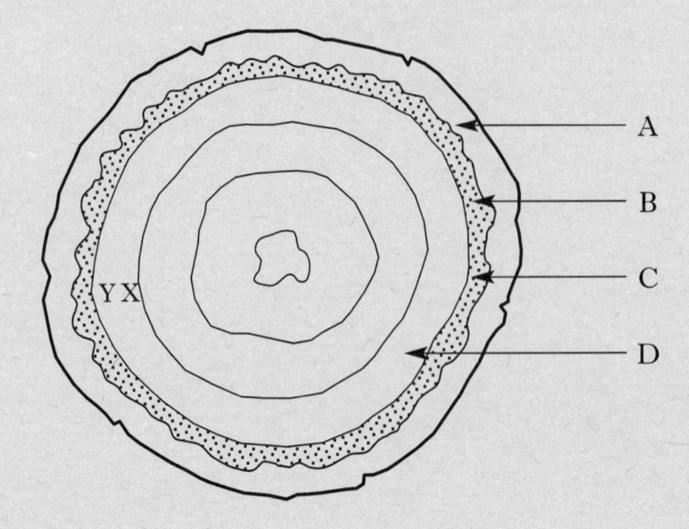

(i) Name the lateral meristem in a woody stem.

________________________________ **1**

(ii) Which letter identifies this tissue?

Letter __________ **1**

(*b*) State whether the average diameter of the vessels in area X would be larger or smaller than those in area Y. Give a reason for your choice.

Average diameter of vessels in area X compared to area Y ________________

Reason ____

____ **1**

DO NOT WRITE IN THIS MARGIN

SECTION C

Marks

Both questions in this section should be attempted.

Note that each question contains a choice.

Questions 1 and 2 should be attempted on the blank pages which follow.

Supplementary sheets, if required, may be obtained from the invigilator.

Labelled diagrams may be used where appropriate.

1. Answer **either** A **or** B.

A. Write notes on each of the following:

(i) the structure of the plasma membrane; **3**

(ii) the structure and function of the cell wall; **3**

(iii) phagocytosis. **4**

(10)

OR

B. Write notes on each of the following:

(i) mRNA synthesis; **5**

(ii) the role of mRNA in protein synthesis. **5**

(10)

In question 2, ONE mark is available for coherence and ONE mark is available for relevance.

2. Answer **either** A **or** B.

A. Give an account of the Jacob-Monod hypothesis of lactose metabolism in Escherichia coli and the part played by genes in the condition of phenylketonuria. **(10)**

OR

B. Give an account of the effects of IAA on plant growth and the role of gibberellic acid in α-amylase induction in barley grains. **(10)**

[*END OF QUESTION PAPER*]

[Turn over

SPACE FOR ANSWERS

ADDITIONAL GRAPH PAPER FOR QUESTION 9(*a*)

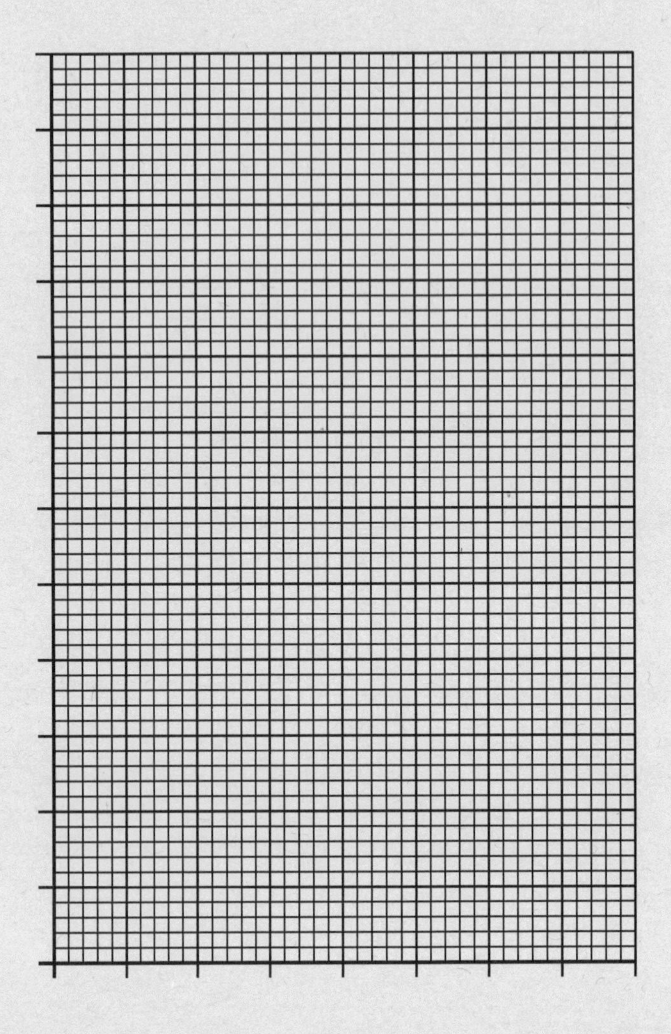

[BLANK PAGE]

FOR OFFICIAL USE

Total for Sections B and C

W007/301

NATIONAL QUALIFICATIONS 2002

WEDNESDAY, 16 JANUARY
9.00 AM – 11.30 AM

BIOLOGY HIGHER

Fill in these boxes and read what is printed below.

Full name of centre

Town

Forename(s)

Surname

Date of birth
Day Month Year

Scottish candidate number

Number of seat

SECTION A—Questions 1–30 (30 marks)

Instructions for completion of Section A are given on page two.

SECTIONS B AND C (100 marks)

1 (a) All questions should be attempted.

(b) It should be noted that in **Section C** questions 1 and 2 each contain a choice.

2 The questions may be answered in any order but all answers are to be written in the spaces provided in this answer book, and must be written clearly and legibly in ink.

3 Additional space for answers and rough work will be found at the end of the book. If further space is required, supplementary sheets may be obtained from the invigilator and should be inserted inside the **front** cover of this book.

4 The numbers of questions must be clearly inserted with any answers written in the additional space.

5 Rough work, if any should be necessary, should be written in this book and then scored through when the fair copy has been written.

6 Before leaving the examination room you must give this book to the invigilator. If you do not, you may lose all the marks for this paper.

SCOTTISH QUALIFICATIONS AUTHORITY

SECTION A

Read carefully

1 Check that the answer sheet provided is for Biology Higher (Section A).

2 Fill in the details required on the answer sheet.

3 In this section a question is answered by indicating the choice A, B, C or D by a stroke made in **ink** in the appropriate place in the answer sheet—see the sample question below.

4 For each question there is only **one** correct answer.

5 Rough working, if required, should be done only on this question paper—or on the rough working sheet provided—**not** on the answer sheet.

6 At the end of the examination the answer sheet for Section A **must** be placed inside the front cover of this answer book.

Sample Question

The apparatus used to determine the energy stored in a foodstuff is a

A respirometer

B calorimeter

C klinostat

D gas burette.

The correct answer is **B**—calorimeter. A **heavy** vertical line should be drawn joining the two dots in the appropriate box in the column headed **B** as shown in the example on the answer sheet.

If, after you have recorded your answer, you decide that you have made an error and wish to make a change, you should cancel the original answer and put a vertical stroke in the box you now consider to be correct. Thus, if you want to change an answer D to an answer B, your answer sheet would look like this:

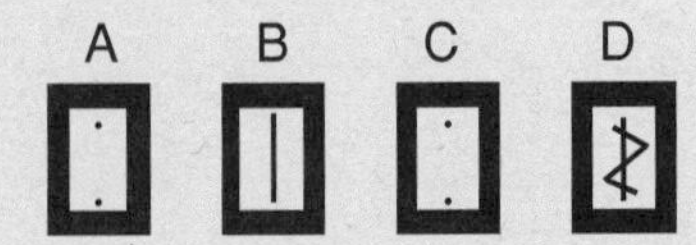

If you want to change back to an answer which has already been scored out, you should enter a tick (✓) to the **right** of the box of your choice, thus:

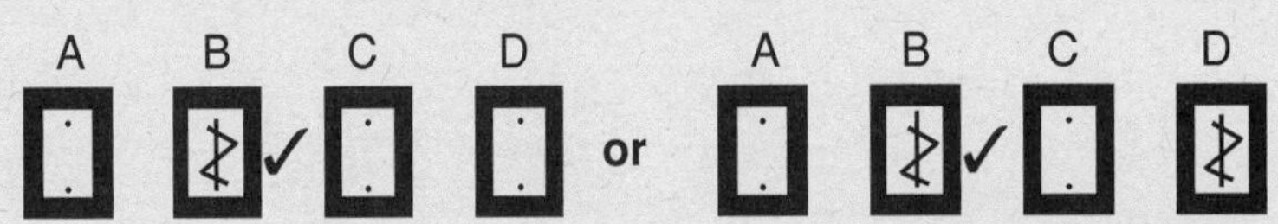

SECTION A

All questions in this section should be attempted.

Answers should be given on the separate answer sheet provided.

1. Which of the following correctly describes a chloroplast?

A Bound by a single membrane with chlorophyll in the grana

B Bound by a single membrane with chlorophyll in the stroma

C Bound by a double membrane with chlorophyll in the grana

D Bound by a double membrane with chlorophyll in the stroma

2. The diagram below shows the appearance of a plant cell which had been placed in an isotonic solution.

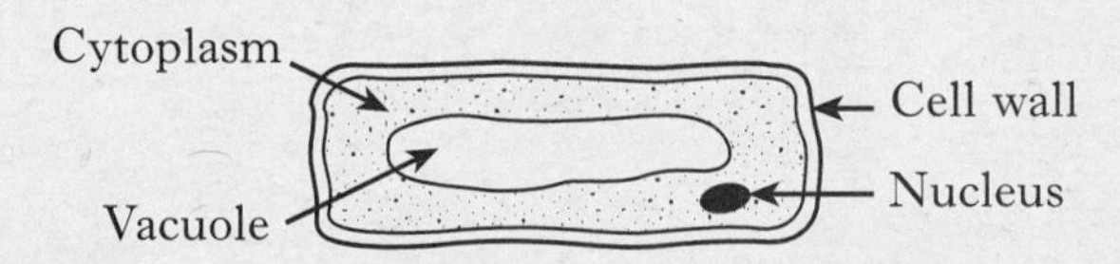

Which of the following diagrams best illustrates the cell after being immersed in a hypertonic solution?

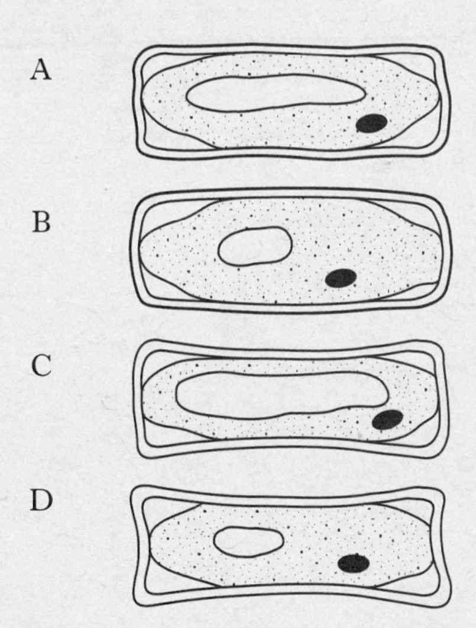

3. Which of the following statements about a young cell wall is **true**?

A It is living.

B It is composed mainly of cellulose.

C It is composed mainly of protein.

D It is selectively permeable.

4. Human red blood cells contain potassium ions at a concentration about 30 times greater than the concentration of potassium ions in the blood plasma. If red blood cells are cooled, potassium ions are lost to the surrounding plasma.

If the cells are warmed again to body temperature, they regain their original concentration of potassium ions.

These movements of potassium ions are explained by

	Outward movement	*Inward movement*
A	diffusion	active transport
B	diffusion	osmosis
C	active transport	diffusion
D	osmosis	active transport

5. In which of the following metabolic pathways is carbon dioxide taken up by ribulose bisphosphate (RuBP)?

A Krebs cycle

B Calvin cycle

C Glycolysis

D Photolysis

6. Assume that, to produce 1 unit of sugar by photosynthesis, each of the plants referred to in the following table had to receive

- 2 units of carbon dioxide
- 2 units of water
- 4 units of light energy.

In which plant was photosynthesis limited by the amount of CO_2 available?

Plant	*Units of CO_2 available to plant*	*Units of H_2O available to plant*	*Units of light available to plant*
A	4	4	7
B	6	8	16
C	4	8	4
D	6	8	11

7. The graphs below show the effect of two injections of an antigen on the formation of an antibody.

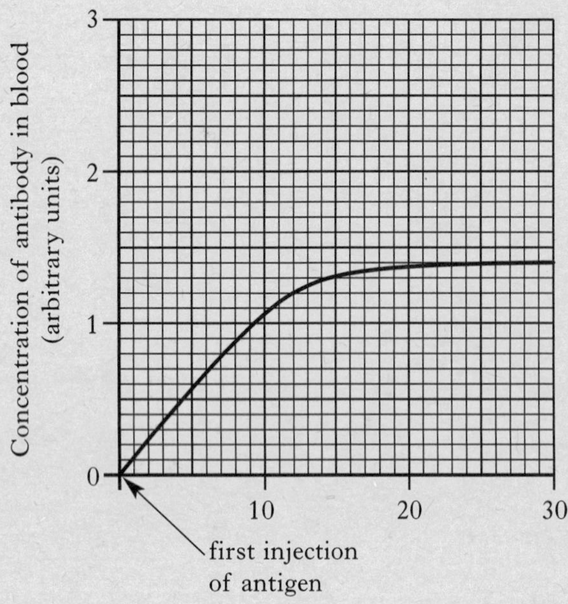

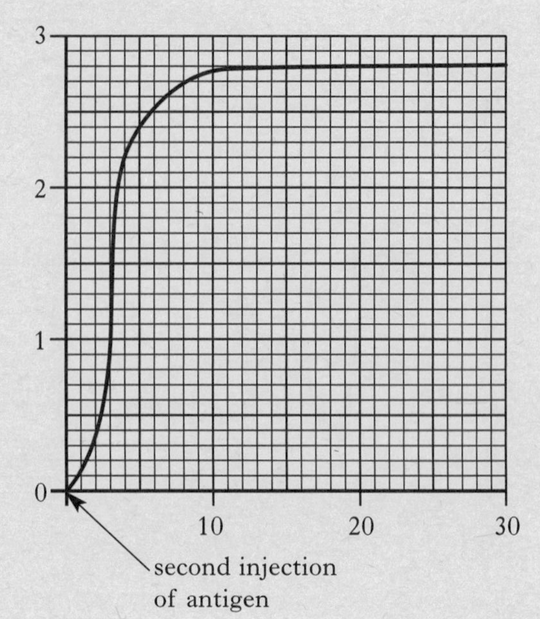

Time in days

The concentration of antibodies is measured 25 days after each injection. The effect of the second injection is to increase the concentration by

A 1%

B 25%

C 50%

D 100%.

---

8. The following information refers to protein synthesis.

tRNA anticodon	*amino acid carried by tRNA*
GUG	Histidine (his)
CGU	Alanine (ala)
GCA	Arginine (arg)
AUG	Tyrosine (tyr)
UAC	Methionine (met)
UGU	Threonine (thr)

What order of amino acids would be synthesised from the base sequence of DNA shown below?

Base sequence of DNA: G C A A T G G T G

A arg – tyr – his

B ala – met – his

C ala – tyr – his

D arg – tyr – thr

9. Which of the following adaptations allow a plant to tolerate grazing by herbivores?

A Thick waxy cuticle

B Low meristems

C Leaves reduced to spines

D Thorny stems

10. Which of the following is **not** a plant response to invasion by other organisms?

A The formation of resin

B The production of nicotine

C The production of antibodies

D The production of tannins

11. The following diagram shows a homologous pair of chromosomes and the loci of 4 genes.

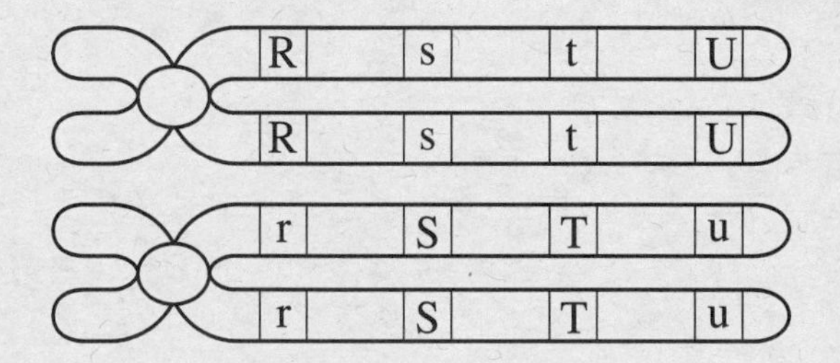

Chiasma formation would occur least often between genes

A r and t

B r and s

C r and U

D s and u.

12. The relative positions of the genes M, N, O and P on a chromosome were determined by the analysis of percentage recombination. The results are shown in the table.

Genes	*Percentage recombination*
M and O	10
N and O	27
N and P	12
M and P	29

The correct order of genes on the chromosome is

A O M P N

B M N O P

C M O N P

D O M N P.

13. Colour blindness is a sex-linked condition. John, who is colour-blind, has the family tree shown below.

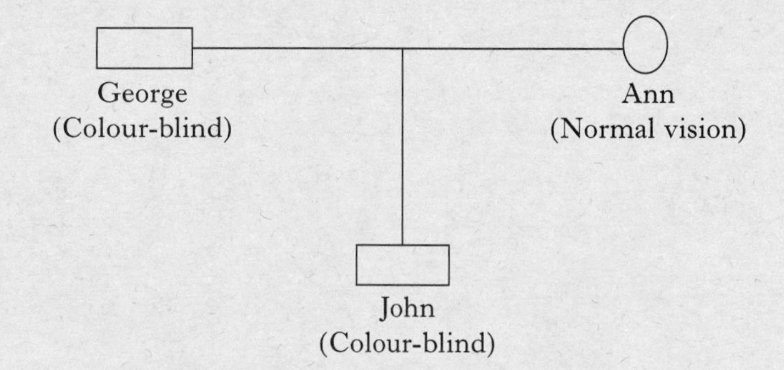

If X^b is a mutant allele and X^B is a normal allele, what were the genotypes of George and Ann's **parents**?

	George's parents		*Ann's parents*	
A	X^BX^b	X^BY	X^BX^B	X^BY
B	X^BX^B	X^bY	X^BX^B	X^BY
C	X^BX^b	X^BY	X^BX^b	X^BY
D	X^BX^B	X^bY	X^BX^B	X^bY

14. A hybrid is created from a cross between two closely related plants whose chromosome numbers are 14 and 28.

Which of the following statements is true for the offspring of this cross?

A Their cells contain 21 chromosomes and they are fertile.

B Their cells contain 21 chromosomes and they are infertile.

C Their cells contain 42 chromosomes and they are fertile.

D Their cells contain 42 chromosomes and they are infertile.

[Turn over

15. A new species of organism is considered to have evolved when a population

A is isolated from the rest of the population by a geographical barrier

B shows increased variation due to mutations

C can no longer interbreed with the rest of the population

D is subjected to increased selection pressures in its habitat.

16. A plant which has a diploid chromosome number 26 was discovered growing in the vicinity of two different species of the same genus which proved to be its parents. The gametes from one parent plant contained 7 chromosomes. What would be the diploid chromosome number of the other plant?

A 12

B 20

C 33

D 40

17. The dark variety of the peppered moth became common in industrial areas of Britain following the increase in the production of soot during the Industrial Revolution.

The increase in the dark form was due to

A dark moths migrating to areas which gave the best camouflage

B a change in the prey species taken by birds

C an increase in the mutation rate

D a change in selection pressure.

18. Below is a list of statements about osmoregulation in fish.

1 Fresh water fish produce a large volume of dilute urine.

2 Salt water bony fish secrete excess salts at the gills.

3 Salt water bony fish have to drink sea water.

Which of the statements are correct?

A 1, 2 and 3

B 1 and 2 only

C 1 and 3 only

D 2 and 3 only

19. The rate of flow of urine of a salmon in fresh water is given as $5 \cdot 0 \text{ cm}^3$/kg of body mass/hour. The volume of urine produced by a $2 \cdot 5$ kg salmon over a period of 5 hours is

A $12 \cdot 5 \text{ cm}^3$

B $25 \cdot 0 \text{ cm}^3$

C $50 \cdot 0 \text{ cm}^3$

D $62 \cdot 5 \text{ cm}^3$.

20. Various factors are involved in causing water to pass through a plant from the soil to the leaves.

Which of the following describes correctly **two** of these factors?

	The water concentration in the soil is	*The water concentration in the xylem of the leaf is*
A	greater than in the root hairs	lower than in the mesophyll cells
B	greater than in the root hairs	greater than in the mesophyll cells
C	lower than in the root hairs	lower than in the mesophyll cells
D	lower than in the root hairs	greater than in the mesophyll cells

21. Which line of the table correctly describes the effect of light and dark on the condition of guard cells and stomatal pores of a green plant?

	Light conditions	*Stomatal pores*	*Guard cells*
A	dark	closed	turgid
B	dark	open	flaccid
C	light	open	flaccid
D	light	open	turgid

22. The diagram below represents the distribution in the intertidal zone of two species of barnacle, X and Y.

The larvae are free-swimming but become attached to rocks where they develop into fixed adults.

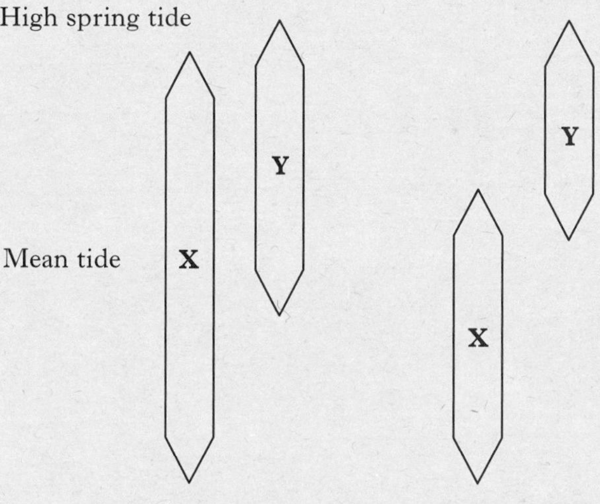

On the basis of the above information, which of the following hypotheses can be made?

A Species X is more tolerant to water loss than species Y.

B There is no competition between the two species.

C Species X is more tolerant of fluctuations in temperature than species Y.

D There is interspecific competition.

23. If a plant has a low compensation point, it may be deduced that

A it requires a high intensity of light for photosynthesis

B it has a high rate of respiration at low intensities of light

C it is able to photosynthesise efficiently at low temperatures

D it is able to photosynthesise efficiently at low intensities of light.

24. The Jacob-Monod model of gene expression involves the following steps.

1 Gene expression

2 Exposure to inducer substances

3 Removal of inhibition

4 Binding to repressor substance

The correct order of these steps is

A 2, 4, 3, 1

B 3, 4, 2, 1

C 4, 1, 2, 3

D 1, 4, 2, 3.

25. The graph shows the growth of an organism over a period of 4 months.

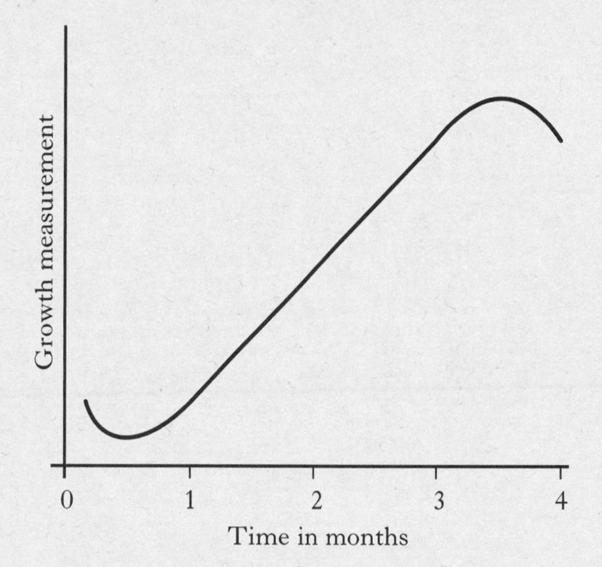

The graph shows changes in the

A mass of an insect

B dry mass of an annual plant

C length of an insect

D length of an annual plant.

[Turn over

26. The following graphs show the relationship between the concentration of the plant growth substance IAA and its promoting or inhibiting effect on the development of certain plant organs.

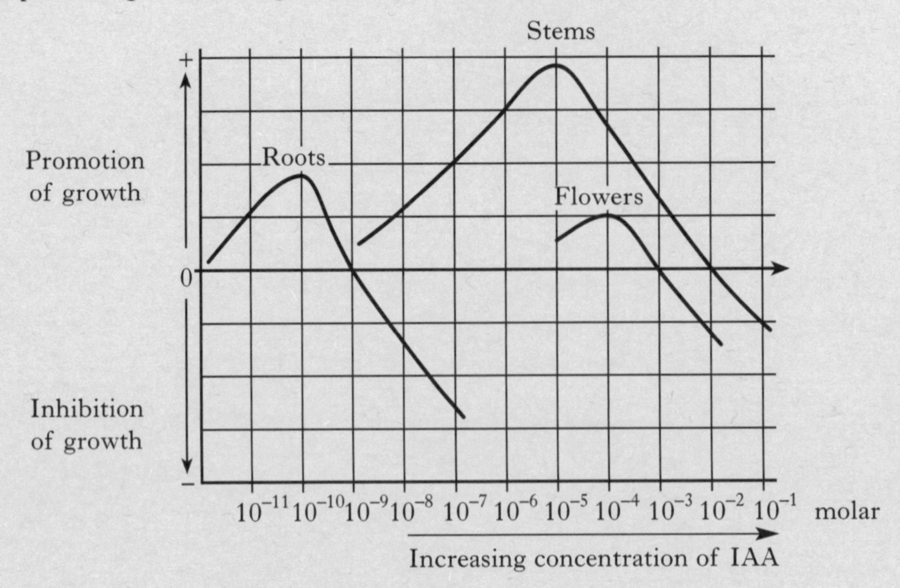

A concentration of 10^{-4} molar solution of IAA will

A promote stem growth but inhibit flower growth

B promote root growth but inhibit stem growth

C promote flower growth and promote stem growth

D inhibit stem growth and inhibit root growth.

27. *Xanthium*, a short-day plant, was exposed to different treatments of dark and light. The diagrams indicate the duration, in hours, of these different treatments.

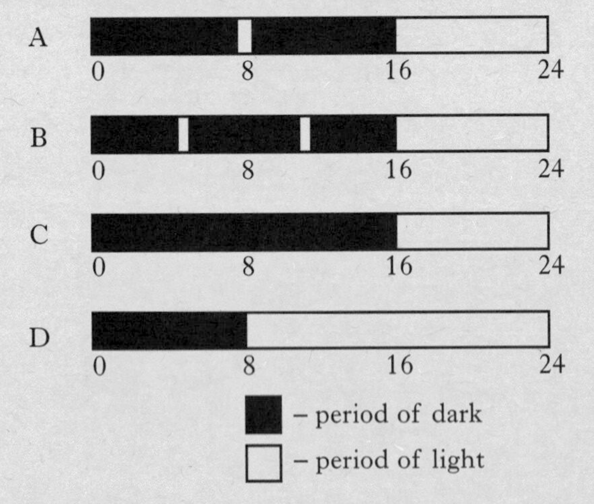

Which of the above treatments would result in flowering?

28. Which of the following is a function of GA (gibberellic acid) but **not** of IAA (indole acetic acid)?

A Breaking of bud dormancy

B Promotion of fruit development

C Stimulation of cell division

D Inhibition of leaf fall

29. During succession in plant communities, a number of changes take place in the ecosystem. Which line of the table correctly describes some of these changes?

	Species diversity	*Biomass*	*Food web complexity*
A	rises	rises	rises
B	rises	falls	rises
C	falls	rises	rises
D	rises	rises	falls

30. Which of the following does **not** occur during succession from a pioneer community of plants to a climax community?

A Soil fertility increases.

B Larger plants replace smaller plants.

C An increasing amount of light reaches ground-dwelling plants.

D Each successive community makes the habitat less favourable for itself.

Candidates are reminded that the answer sheet MUST be returned INSIDE the front cover of this answer book.

[Turn over for Section B on *Page ten*]

Marks

SECTION B

All questions in this section should be attempted.

1. The diagram below represents some of the structures present in an animal cell.

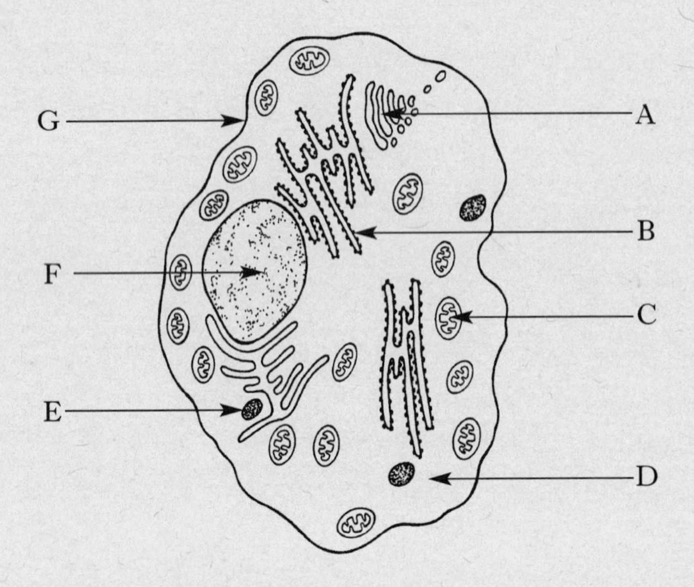

Complete the table by using a letter from the diagram to show which structure carries out each function.

(Each letter may be used **once**, **more than once** or **not at all**.)

Function	*Letter*
mRNA synthesis	
Translation of mRNA	
Contains enzymes for use in phagocytosis	
Processing and packaging for secretion	
Aerobic ATP production	
Selective ion uptake	

3

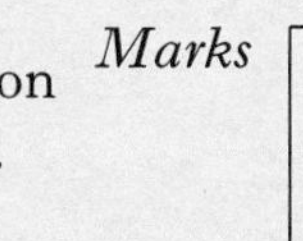

Marks

2. (*a*) The graph below shows the effects of increasing carbon dioxide concentration on the rate of photosynthesis at different light intensities and temperatures.

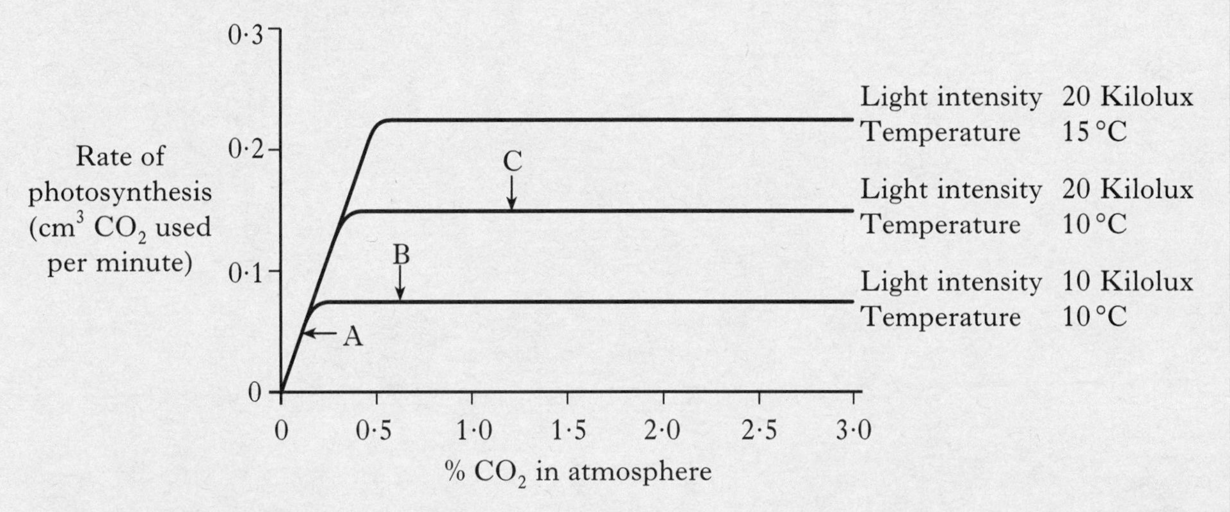

Use the information in the graph to complete the table below.

Tick (✓) **one** box in each row to indicate the factor which is limiting the rate of photosynthesis at points A, B and C.

Graph point	*Light intensity*	*Temperature*	*Carbon dioxide concentration*
A			
B			
C			

2

[Turn over

Marks

2. (continued)

(*b*) Diagram 1 below shows a leafy shoot. Sugars manufactured in leaf A are transported to other parts of the shoot. Diagram 2 shows the distribution of these sugars 18 hours later.

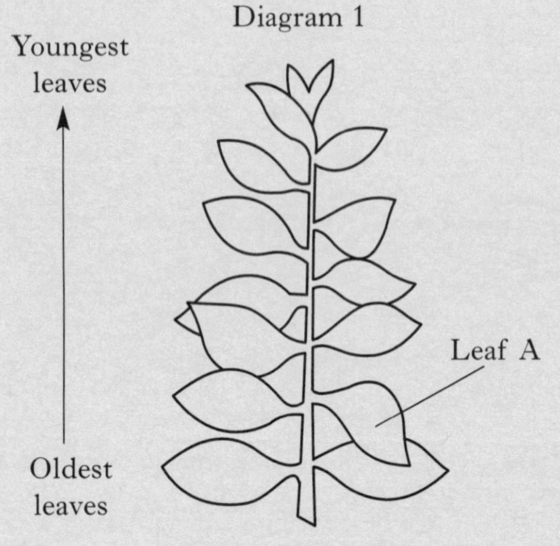

Leafy shoot at start

Diagram 2

Shading shows where sugars made in leaf A were present after 18 hours

From the results, suggest **two** conclusions that can be drawn about the transport of sugars in the leafy shoot over the 18 hour period.

Conclusion 1 ____

____

____ **1**

Conclusion 2 ____

____

____ **1**

DO NOT WRITE IN THIS MARGIN

Marks

3. The diagram below represents stages in respiration in yeast.

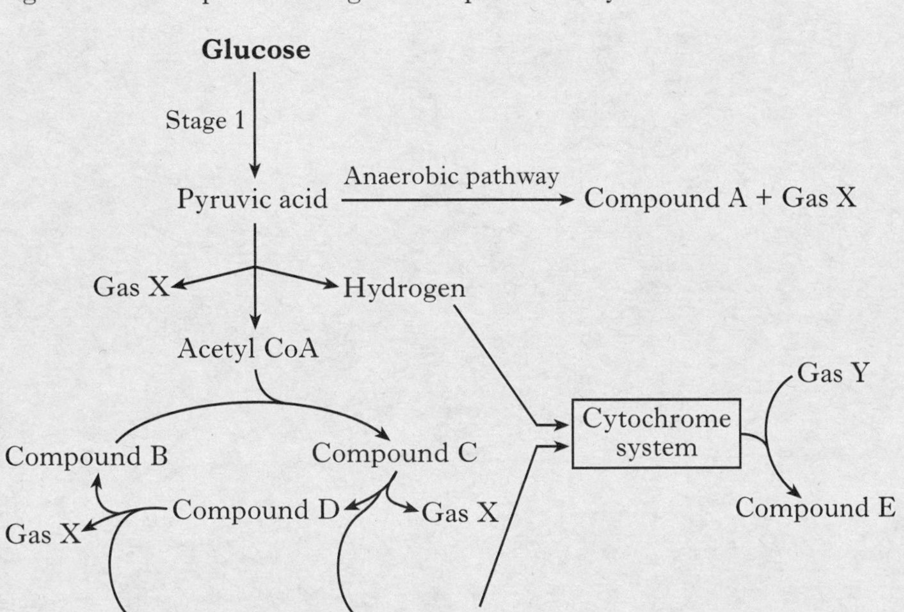

(*a*) Complete the table below by inserting a letter from the diagram to identify a compound which has the number of carbon atoms shown.

Number of carbon atoms	*Compound*
6	
4	
2	

2

(*b*) Name Stage 1 and state its location in the yeast cell.

Name ______________________

Location ______________________ **1**

(*c*) Apart from glucose, name **two** other substances which must be present for Stage 1 to take place.

______________________ and ______________________ **1**

(*d*) (i) Name the carrier which transports hydrogen to the cytochrome system.

______________________ **1**

(ii) Name gas Y and compound E.

Gas Y ______________________

Compound E ______________________ **1**

(*e*) Name a product of aerobic respiration which is **not** shown in the diagram.

______________________ **1**

DO NOT WRITE IN THIS MARGIN

Marks

4. The diagram below represents replication of part of a DNA molecule.

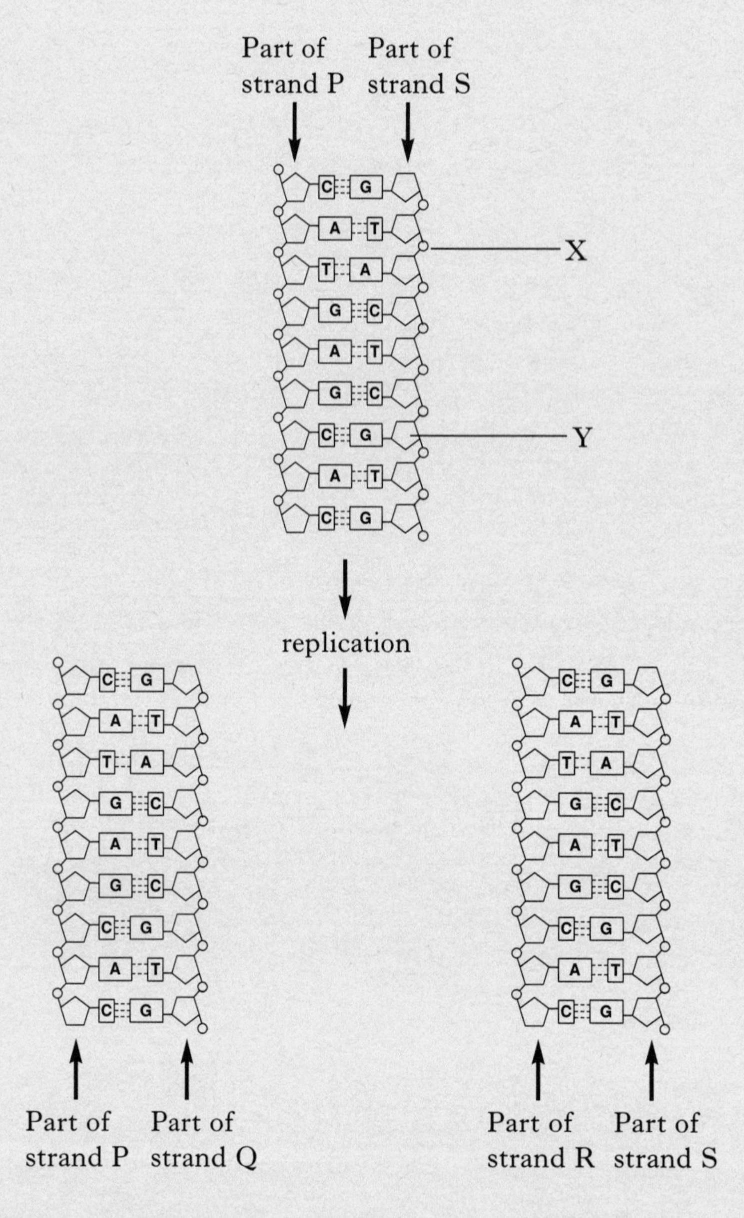

(*a*) Name the components labelled X and Y on the diagram.

X ______________________________

Y ______________________________ **1**

(*b*) Apart from length, give **one** structural feature of a DNA molecule which is **not** shown in the diagram.

____ **1**

DO NOT WRITE IN THIS MARGIN

Marks

4. (continued)

(*c*) Name the bonds shown in the diagram which must be broken at the start of replication.

______________________________ **1**

(*d*) The full length of DNA strand P has 18% adenine bases and 26% cytosine bases.

(i) Calculate the combined percentage of thymine and guanine bases in strand P.

Space for calculation

Percentage ________ % **1**

(ii) Calculate the percentage of guanine bases in strand S.

Space for calculation

Percentage ________ % **1**

(*e*) The base sequence of the nucleotides on the part of strand P in the diagram is CATGAGCAC.

(i) Write a possible base sequence for this part of strand P in each of the following mutations.

1 Insertion ______________________________ **1**

2 A single substitution ______________________________ **1**

(ii) Other than chemical substances, name **one** mutagenic agent that can cause such changes to the base sequence of DNA.

______________________________ **1**

[Turn over

5. The diagram below shows a respirometer which was used to measure the rate of respiration of various living materials. The investigation was carried out at 20 °C.

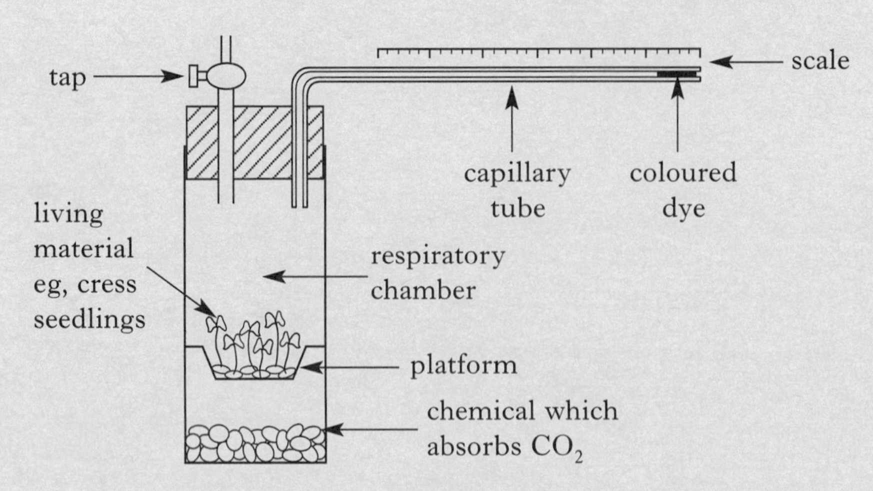

The living material was placed on the platform in the respiratory chamber. The tap was left open for 10 minutes. The tap was then closed and coloured dye was introduced into the end of the capillary tube. The capillary tube had a cross-sectional area of 3 mm². The rate of respiration is expressed as mm³ of oxygen used per gram of living material per minute and can be calculated by using the following formula.

$$\text{Rate of respiration} = \frac{\text{Cross - sectional area of capillary tube (mm}^2\text{)} \times \text{distance moved by coloured dye (mm)} \times 60}{\text{Time taken for the coloured dye to move the distance (s)} \times \text{mass of living material (g)}}$$

The table below shows the results of the investigation.

Living material	*Mass of living material* (g)	*Distance moved by coloured dye* (mm)	*Time taken* (s)	*Rate of respiration* (mm^3 O_2/g/min)
Cress seedlings	30	25	60	2·5
Apple tissue	37·5	25	120	1·0
Fresh liver tissue	60	50	15	10·0
Earthworm	10	15	45	

DO NOT WRITE IN THIS MARGIN

Marks

(*a*) **Complete the table** by calculating the rate of respiration of an earthworm.

Space for calculation

1

(*b*) Different masses of living tissue were used in each experiment.

Explain how the results still allow a valid comparison of the rates of respiration.

____

____ **1**

DO NOT WRITE IN THIS MARGIN

Marks

5. (continued)

(*c*) With some living tissue it was found that over a 10 minute period the distance moved by the dye was too little to be measured accurately.

How could the apparatus be modified to overcome this problem?

1

(*d*) To improve the reliability of the results for cress seedlings, the experiment was replicated six times. The results are shown in the table below.

Trial	1	2	3	4	5	6
Results (mm^3 O_2/g/min)	2·52	2·53	2·48	2·49	0·20	2·51

The results of trial 5 differ from the others. Suggest **one** possible source of error that could account for this result.

1

(*e*) The experiment with cress seedlings was carried out in darkness. Explain why the results would have been invalid if the experiment had been carried out in the light.

2

[Turn over

DO NOT WRITE IN THIS MARGIN

Marks

5. (continued)

(*f*) In a further investigation the rate of respiration of fresh liver tissue was measured over a range of temperatures. The results are shown in the table below.

Temperature (°C)	10	15	30	35	40	45
Rate of respiration (mm^3 O_2/g/min)	2·0	4·0	9·5	13·0	18·5	15·5

Using the **full area** of the grid below, plot a line graph of rate of respiration against temperature.

(Additional graph paper, if required, can be found on page 32.)

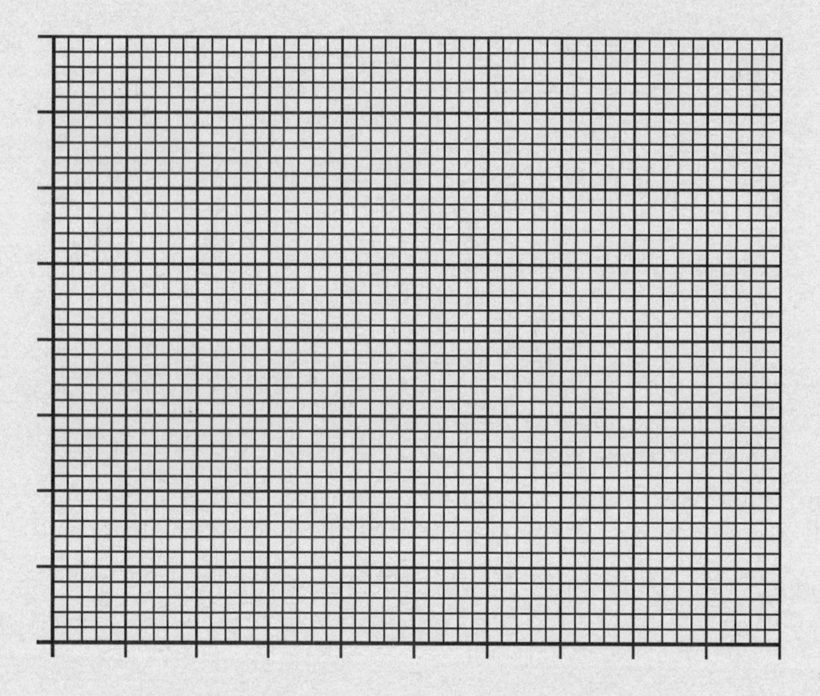

2

DO NOT WRITE IN THIS MARGIN

Marks

6. (*a*) The table below shows the characteristics of the leaves of two plant species.

Species	*Cuticle*	*Leaf hairs*
Phaseolus	Thin	Absent
Pelargonium	Thick	Present

The graph below compares transpiration from these plants.

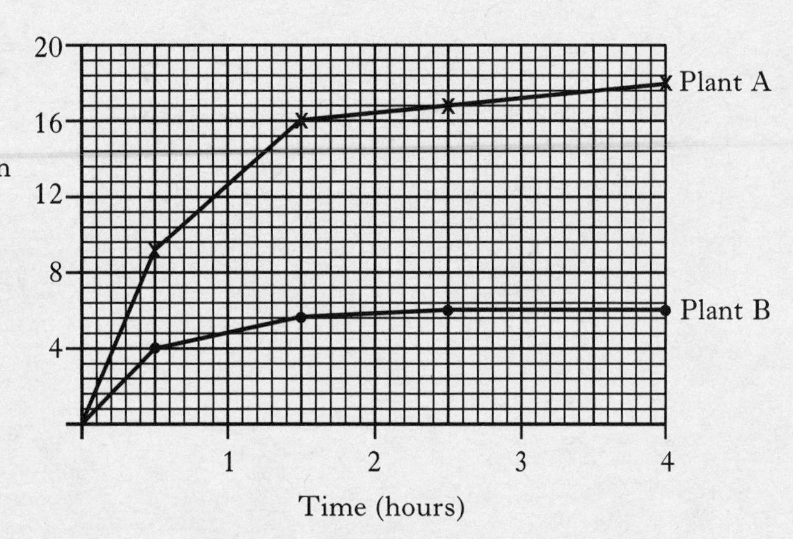

From the graph, identify the plant, A or B which is *Pelargonium* and justify your choice.

Plant __________

Justification ____

____

____ **1**

(*b*) Environmental changes have an effect on the rate of transpiration of plants such as *Phaseolus sp* and *Pelargonium sp*.

Complete the table below by using the words **increase**, **decrease** or **no change** to show the effect.

Environmental change	*Effect on rate of transpiration*
Decreased humidity	
Decreased temperature	
Light to dark	
Increased wind speed	

2

Marks

6. (continued)

(*c*) The diagram below shows a transverse section through the stem of a hydrophyte.

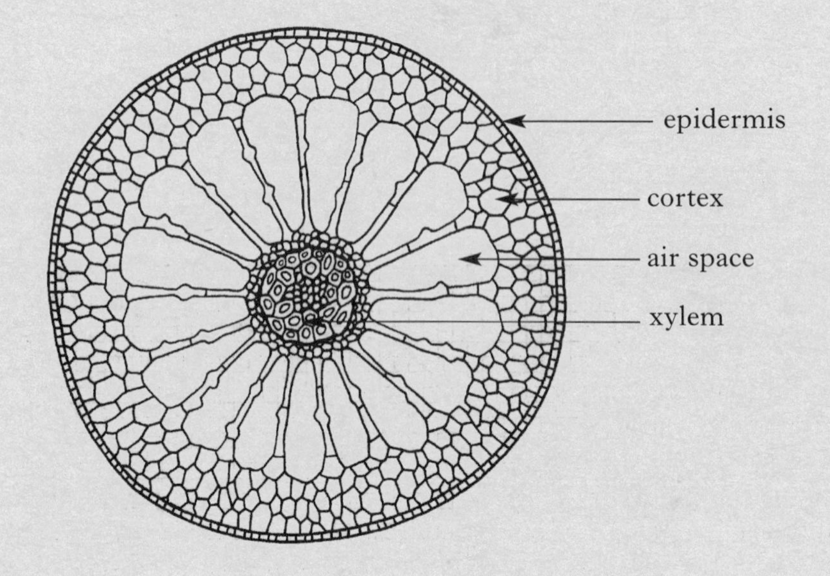

(i) What is the function of the air spaces?

____

____ **1**

(ii) State the benefit of the xylem tissue being situated at the centre of the stem.

____

____ **1**

Marks

7. The diagram below represents an animal cell which is dividing.

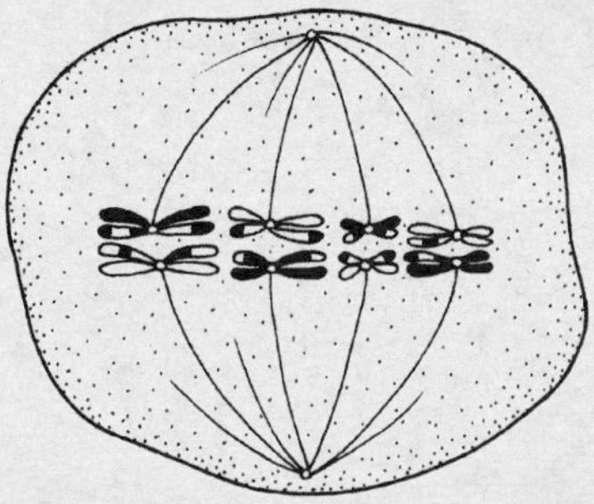

(*a*) Select **two** features shown in the diagram which show that the cell is dividing by meiosis.

Feature 1 ____________________

____________________ **1**

Feature 2 ____________________

____________________ **1**

(*b*) The following diagram shows the same cell at a later stage in meiosis.

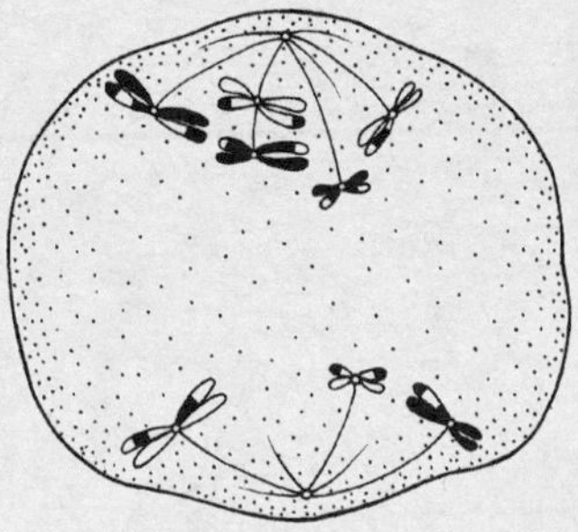

(i) What type of chromosome mutation is shown in the diagram?

____________________ **1**

(ii) State the chromosome numbers which are present in the gametes formed from this cell.

Chromosome numbers __________ and __________ **1**

(*c*) A human liver cell contains 6·0 units of DNA.

How many units of DNA are present in the following human cells?

1 Ovum __________ units

2 Gamete mother cell __________ units **1**

[Turn over

Marks

8. Feather colour in budgerigars is controlled by two genes located on different chromosomes. Each gene has two alleles: A is dominant to a, and B is dominant to b.

Information on genotypes and their associated phenotypes is shown in the table below.

Genotypes	*Phenotypes*
AAbb, Aabb	yellow
aaBB, aaBb	blue
aabb	white
Alleles A and B both present	green

A male of genotype aaBb was crossed with a female of genotype AaBb.

(*a*) State the phenotype of each bird.

male ______________________ female ______________________ **1**

(*b*) Complete the table below by:

1 inserting the genotypes of the male and female gametes **1**

2 showing the possible genotypes of the offspring. **1**

Gametes		*Female*			
Male					

(*c*) In the spaces below, show the expected phenotype ratio of the offspring.

________ green : ________ blue : ________ yellow : ________ white **1**

DO NOT WRITE IN THIS MARGIN

Marks

9. The diagram below shows a transverse section through the stem of a tree that was cut down in November 2001.

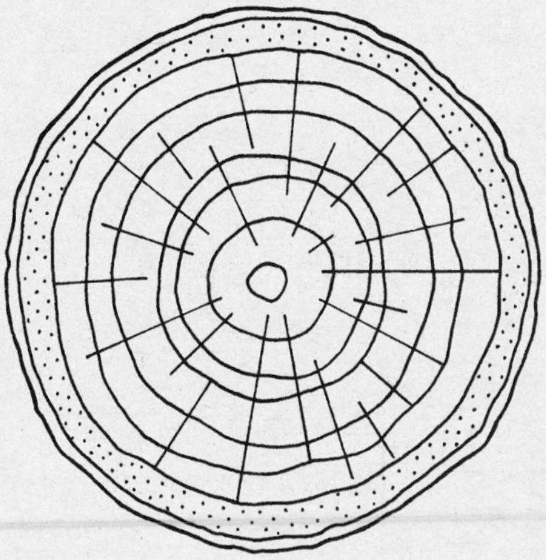

(*a*) **Insert an X** within the annual ring that was formed in 1999. **1**

(*b*) One of the annual rings is much narrower than the others.

Explain how heavy infestation by leaf-eating caterpillars may account for this observation.

____

____

____ **2**

(*c*) Describe **one** way in which the structure of a spring xylem vessel differs from that of a summer xylem vessel.

____

____ **1**

(*d*) Name the meristem that is responsible for the formation of the new cells that differentiate into xylem vessels.

______________________ **1**

(*e*) How can the control of cell differentiation be explained in terms of gene activity?

____

____ **1**

[Turn over

Marks

10. The graph below shows changes in the α-amylase concentration and the starch content of a barley grain during early growth and development.

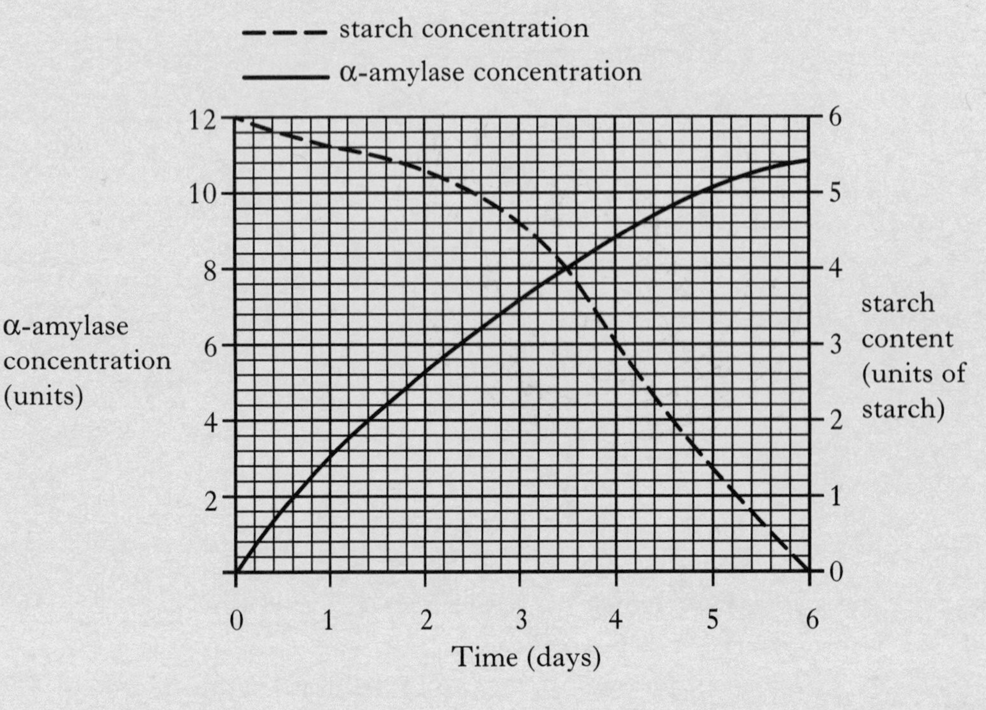

(*a*) (i) After how many days is the starch content of the barley grains decreased by 50%?

____________________ **1**

(ii) Account for the increasing rate of breakdown of starch as shown in the graph.

____

____________________ **1**

(*b*) In which tissue of the barley grain is α-amylase synthesised?

____________________ **1**

(*c*) Name the following.

1 The plant growth substance that initiates the synthesis of α-amylase

____________________ **1**

2 The site of production of this plant growth substance within the barley grain

____________________ **1**

DO NOT WRITE IN THIS MARGIN

Marks

11. The diagram below outlines the role of a gland and three hormones which influence growth and development in humans.

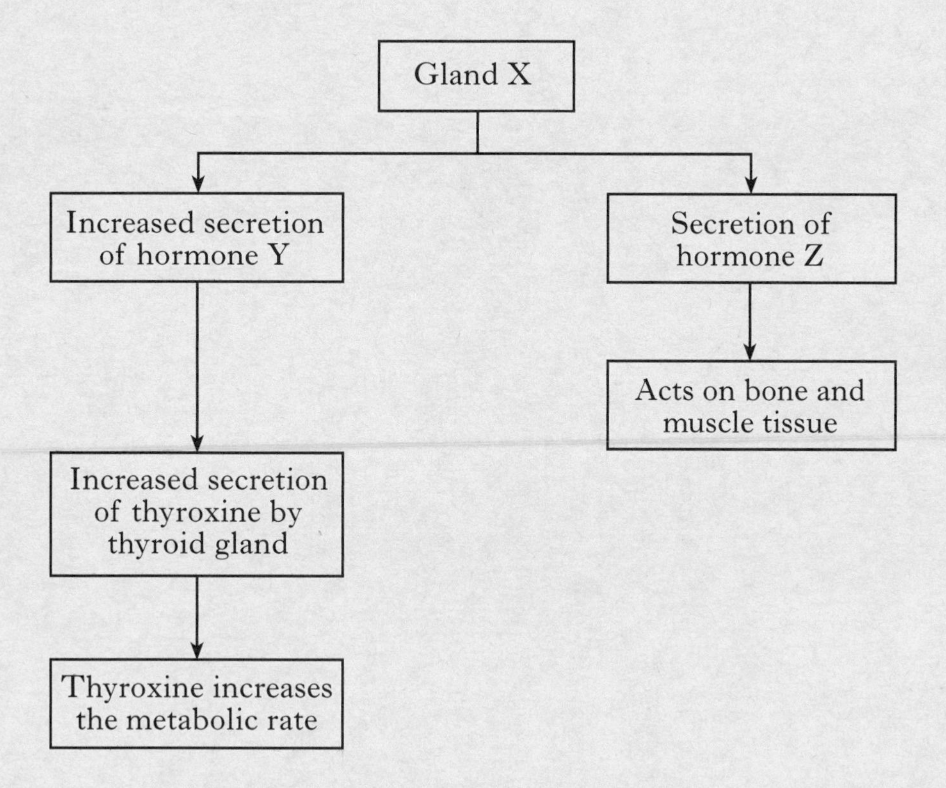

(*a*) Name gland X and hormones Y and Z.

Gland X ____________________

Hormone Y ____________________ Hormone Z ____________________ **2**

(*b*) The control of thyroxine secretion is an example of negative feedback control. Given this information, predict the effect of an increase in thyroxine on gland X.

____

____ **1**

(*c*) Name a chemical element and a vitamin needed for the growth and development of bone.

Chemical element ____________________

Vitamin ____________________ **1**

[Turn over

DO NOT WRITE IN THIS MARGIN

Marks

12. Plants require the chemical elements nitrogen, phosphorus, magnesium and potassium for normal growth and development.

The boxes below refer to:

1 the appearance of plants that results from deficiencies in these elements

2 the importance of these elements in normal growth and development of plants.

A	B	C
Leaf bases appear red	Overall growth of the plant is reduced	Roots show increased growth in length
D	E	F
Leaves are chlorotic	Essential in the formation of chlorophyll	Essential for membrane transport
G	H	I
Essential in the formation of ATP	Essential in the formation of DNA and RNA	Essential for the synthesis of proteins

(*a*) Which **two** boxes would describe the appearance of plants which were grown in soil deficient in phosphorus?

Letters ______________ and ______________ **1**

(*b*) Which **three** boxes give reasons why nitrogen is essential for normal growth and development in plants?

Letters ______________ , ______________ and ______________ **2**

(*c*) A deficiency in different elements can result in the same abnormal growth and development.

Name **two** elements which, if deficient, cause leaves to appear chlorotic (box D).

Elements ______________________ and ______________________ **1**

(*d*) Give an account of the link between boxes B and H.

____

____

____ **2**

DO NOT WRITE IN THIS MARGIN

Marks

13. The table below shows the relationship between the concentration of lead in the placenta and the average birth mass of human babies.

Range of concentration of lead in placenta (units)	*Average birth mass* (kg)
25 – 29	2·32
20 – 24	2·87
15 – 19	3·40
10 – 14	3·74
5 – 9	4·40

(*a*) (i) Describe the relationship between the concentration of placental lead and average birth mass. **1**

(ii) State the effect of lead on cell functions. **1**

(*b*) (i) Describe the effect of thalidomide on fetal development. **1**

(ii) Name **two** drugs which have the effect of causing a decrease in the expected birth weight.

Drug 1 ____________________ Drug 2 ____________________ **1**

[Turn over

Marks

14. During pharmaceutical trials for a new medicine, a healthy subject volunteered to drink 1 litre of water.

Samples of urine were taken over a 3 hour period as it flowed from the kidney to the bladder.

These samples were used to determine the rate of urine production and the salt concentration of the urine.

The results are shown in Graph 1 below.

Graph 1

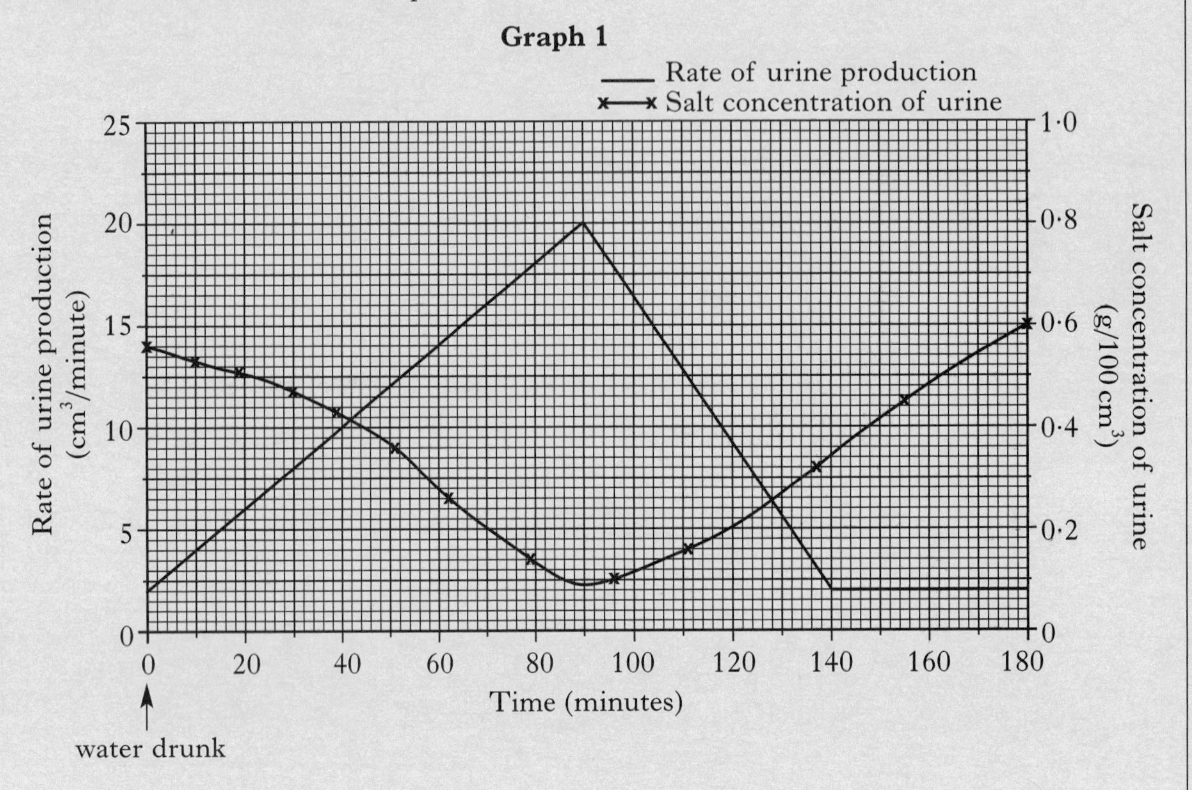

(*a*) From **Graph 1**, how long did it take, after drinking the water, for the rate of urine production to return to the initial value?

Time ________________ minutes **1**

(*b*) From **Graph 1**, describe the relationship between the rate of urine production and salt concentration in the urine over the 3 hour period.

____

____

____

________________________________ **2**

(*c*) From **Graph 1**, calculate the decrease in the salt concentration of the urine over the first 90 minutes.

Space for calculation

______________ g/100 cm^3 **1**

DO NOT WRITE IN THIS MARGIN

14. (continued) *Marks*

In a further investigation, 10 mg of the medicine was administered to the volunteer who then drank 1 litre of water.

The results are shown in **Graph 2**.

Graph 2

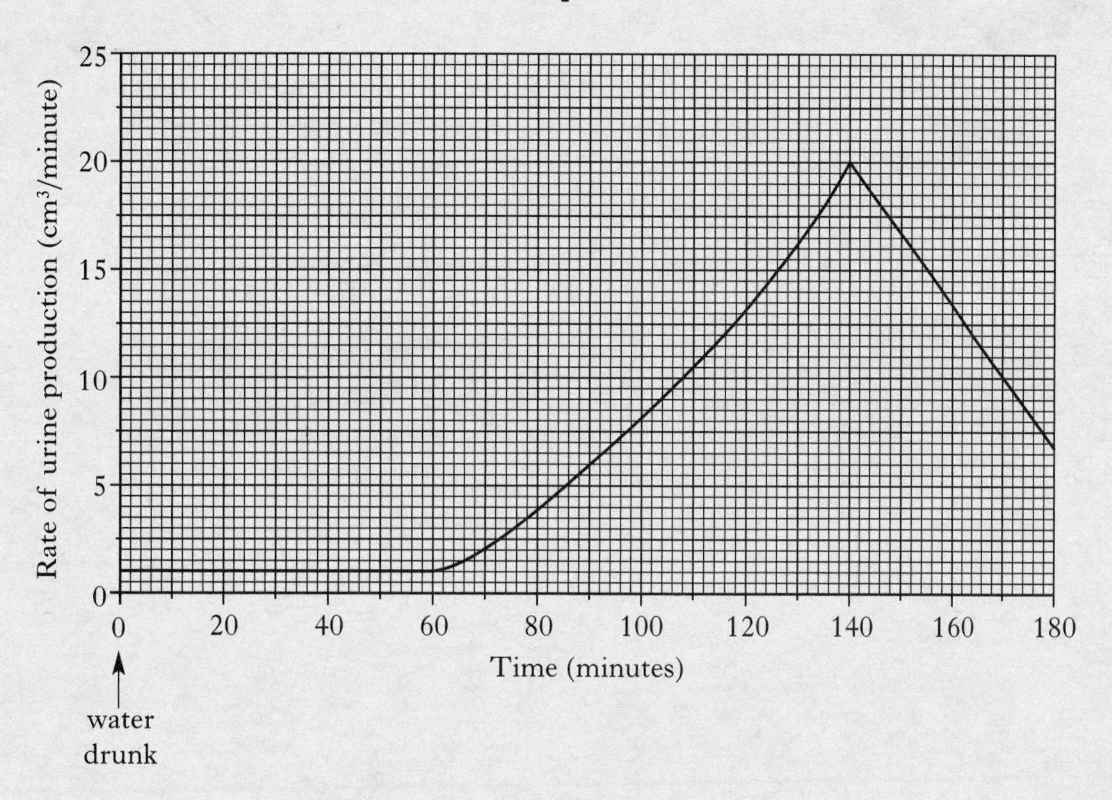

(*d*) From **Graphs 1** and **2**, calculate the difference between the rates of production of urine 90 minutes after the start.

Space for calculation

____________ cm^3/minute **1**

(*e*) From **Graphs 1** and **2**, explain how the evidence supports the statement that the medicine affects the rate of water reabsorption in the kidney.

____

____

____ **2**

(*f*) Predict how the result of the trial may have differed if 20 mg of the medicine had been administered.

____

____ **1**

DO NOT WRITE IN THIS MARGIN

Marks

15. Various factors have an effect on the size of a population.

The flowchart outlines how population density may be regulated.

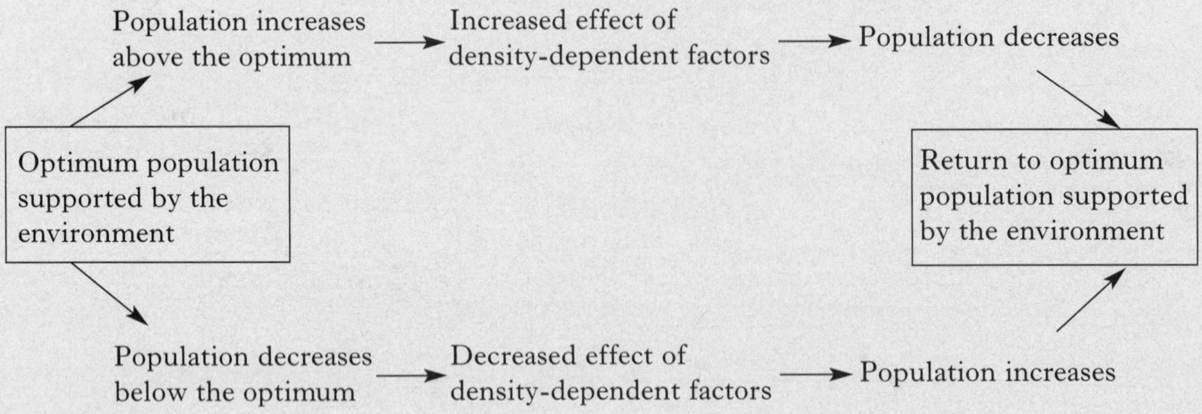

Population increases above the optimum → Increased effect of density-dependent factors → Population decreases → Return to optimum population supported by the environment

Optimum population supported by the environment → Population decreases below the optimum → Decreased effect of density-dependent factors → Population increases → Return to optimum population supported by the environment

(*a*) Competition for food between members of the same species has a density-dependent effect on a population. What term describes competition between members of the same species?

______________________________ **1**

(*b*) Other than competition for food between members of the same species, name **two** other density-dependent factors that affect population size and explain how each factor has its effect.

1 Factor ______________

Explanation of effect ______________________________

____

______________________________ **1**

2 Factor ______________

Explanation of effect ______________________________

____

______________________________ **1**

DO NOT WRITE IN THIS MARGIN

SECTION C

Both questions in this section should be attempted.

Note that each question contains a choice.

Questions 1 and 2 should be attempted on the blank pages which follow.

Supplementary sheets, if required, may be obtained from the invigilator.

Labelled diagrams may be used where appropriate.

Marks

1. Answer **either** A **or** B.

A. Write notes on each of the following:

(i) photosynthetic pigments; **5**

(ii) the light-dependent stage of photosynthesis. **5**

(10)

OR

B. Write notes on each of the following:

(i) invasion of cells by viruses and the production of more viruses; **5**

(ii) cellular defence mechanisms in animals. **5**

(10)

In question 2, ONE mark is available for coherence and ONE mark is available for relevance.

2. Answer **either** A **or** B.

A. Give an account of the importance of the physiological and behavioural adaptations shown by the desert rat in water conservation. **(10)**

OR

B. Give an account of artificial selection with reference to selective breeding and somatic fusion. **(10)**

[*END OF QUESTION PAPER*]

[Turn over

DO NOT WRITE IN THIS MARGIN

SPACE FOR ANSWERS

ADDITIONAL GRAPH PAPER FOR QUESTION 5(*f*)

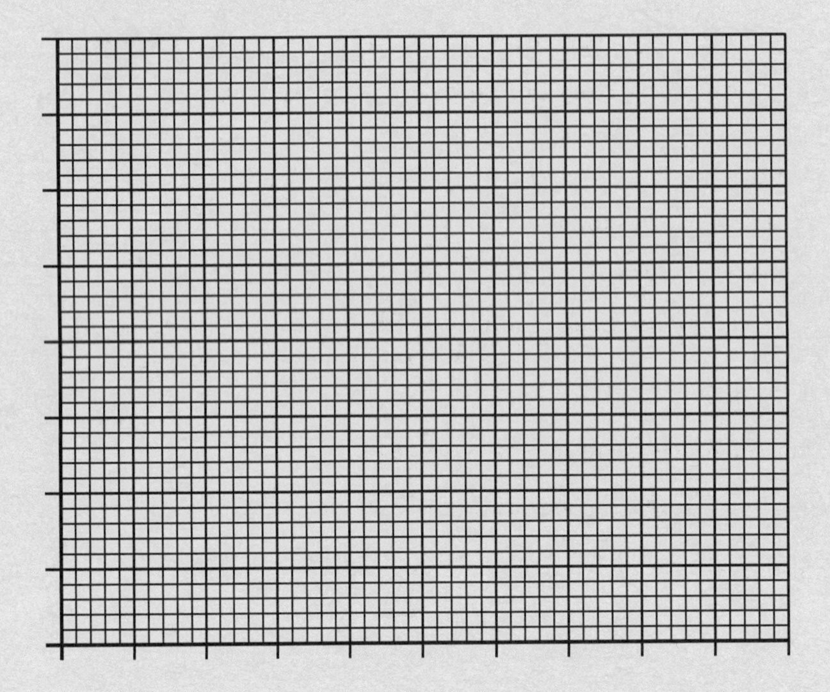

[BLANK PAGE]

FOR OFFICIAL USE

Total for Sections B and C

X007/301

NATIONAL QUALIFICATIONS 2003

MONDAY, 26 MAY
1.00 PM – 3.30 PM

BIOLOGY
HIGHER

Fill in these boxes and read what is printed below.

Full name of centre

Town

Forename(s)

Surname

Date of birth
Day Month Year

Scottish candidate number

Number of seat

SECTION A—Questions 1–30 (30 marks)

Instructions for completion of Section A are given on page two.

SECTIONS B AND C (100 marks)

1 (a) All questions should be attempted.

(b) It should be noted that in **Section C** questions 1 and 2 each contain a choice.

2 The questions may be answered in any order but all answers are to be written in the spaces provided in this answer book, and must be written clearly and legibly in ink.

3 Additional space for answers and rough work will be found at the end of the book. If further space is required, supplementary sheets may be obtained from the invigilator and should be inserted inside the **front** cover of this book.

4 The numbers of questions must be clearly inserted with any answers written in the additional space.

5 Rough work, if any should be necessary, should be written in this book and then scored through when the fair copy has been written.

6 Before leaving the examination room you must give this book to the invigilator. If you do not, you may lose all the marks for this paper.

SECTION A

Read carefully

1 Check that the answer sheet provided is for Biology Higher (Section A).

2 Fill in the details required on the answer sheet.

3 In this section a question is answered by indicating the choice A, B, C or D by a stroke made in **ink** in the appropriate place in the answer sheet—see the sample question below.

4 For each question there is only **one** correct answer.

5 Rough working, if required, should be done only on this question paper—or on the rough working sheet provided—**not** on the answer sheet.

6 At the end of the examination the answer sheet for Section A **must** be placed inside the front cover of this answer book.

Sample Question

The apparatus used to determine the energy stored in a foodstuff is a

A respirometer

B calorimeter

C klinostat

D gas burette.

The correct answer is **B**—calorimeter. A **heavy** vertical line should be drawn joining the two dots in the appropriate box in the column headed **B** as shown in the example on the answer sheet.

If, after you have recorded your answer, you decide that you have made an error and wish to make a change, you should cancel the original answer and put a vertical stroke in the box you now consider to be correct. Thus, if you want to change an answer D to an answer B, your answer sheet would look like this:

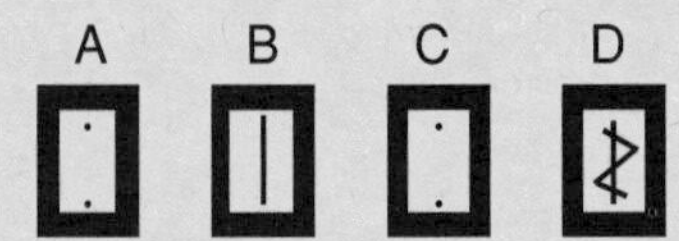

If you want to change back to an answer which has already been scored out, you should enter a tick (✓) to the **right** of the box of your choice, thus:

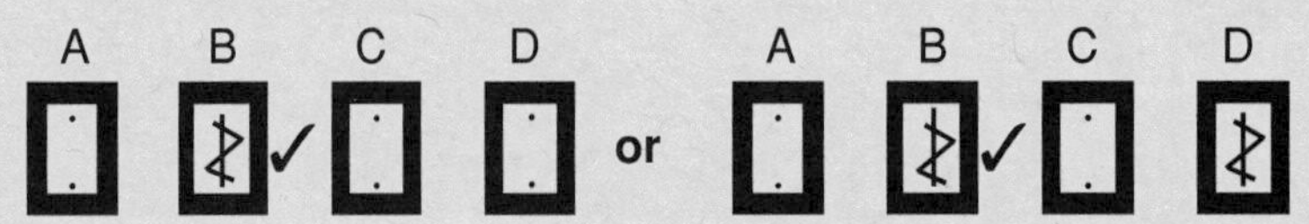

SECTION A

All questions in this section should be attempted.

Answers should be given on the separate answer sheet provided.

1. The table below shows the concentrations of three ions found in sea water and in the sap of the cells of a seaweed.

	Ion concentrations ($mg\,l^{-1}$)		
	potassium	*sodium*	*chloride*
sea water	0·01	0·55	0·61
cell sap	0·57	0·04	0·60

Which of the following statements is supported by the data in the table?

A Potassium and sodium ions are taken into the cell by active transport.

B Potassium and chloride ions are removed from the cell by diffusion.

C Sodium ions are removed from the cell by active transport.

D Chloride and sodium ions are removed from the cell by diffusion.

2. A piece of muscle was cut into three strips, X, Y and Z, and treated as described in the table.

Their final lengths were then measured.

Muscle strip	*Solution added to muscle*	*Muscle length* (mm)	
		Start	*After 10 minutes*
X	1% glucose	50	50
Y	1% ATP	50	45
Z	1% ATP boiled and cooled	50	46

From the data it may be deduced that

A ATP is not an enzyme

B muscles contain many mitochondria

C muscles synthesise ATP in the absence of glucose

D muscles do not use glucose as a source of energy.

3. DNA controls the activities of a cell by coding for the production of

A proteins

B carbohydrates

C amino acids

D bases.

4. The diagram below shows part of a DNA molecule during replication. Bases are represented by numbers and letters.

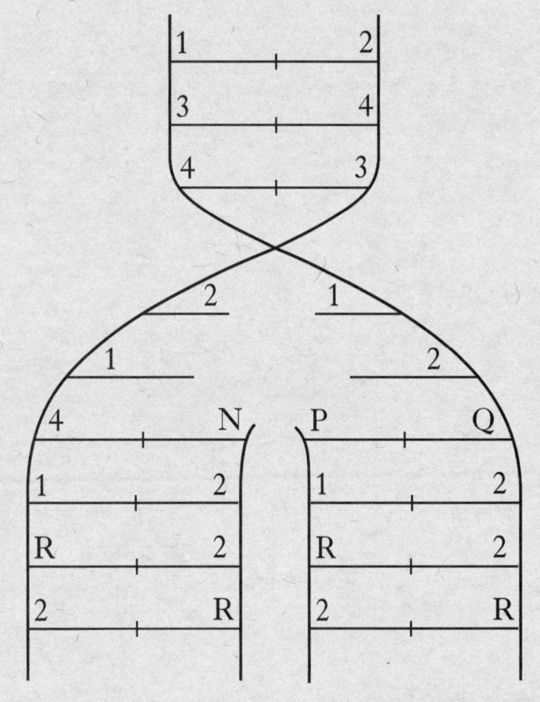

If 1 represents adenine and 3 represents cytosine, which line in the table identifies correctly the bases represented by the letters N, P, Q and R?

	N	*P*	*Q*	*R*
A	guanine	cytosine	guanine	thymine
B	cytosine	guanine	cytosine	adenine
C	guanine	cytosine	cytosine	adenine
D	cytosine	guanine	guanine	adenine

[Turn over

5. The table below contains statements which may be TRUE or FALSE concerning DNA replication and mRNA synthesis.

Which line in the table is correct?

	Statement	*DNA replication*	*mRNA synthesis*
A	Occurs in the nucleus	TRUE	FALSE
B	Involved in protein synthesis	TRUE	TRUE
C	Requires free nucleotides	TRUE	FALSE
D	Involves complementary base pairing	TRUE	TRUE

6. A fragment of DNA was found to have 60 guanine bases and 30 adenine bases. What is the total number of deoxyribose sugar molecules in this fragment?

A 30

B 45

C 90

D 180

7. The diagram represents part of a molecule of DNA on which a molecule of RNA is being synthesised.

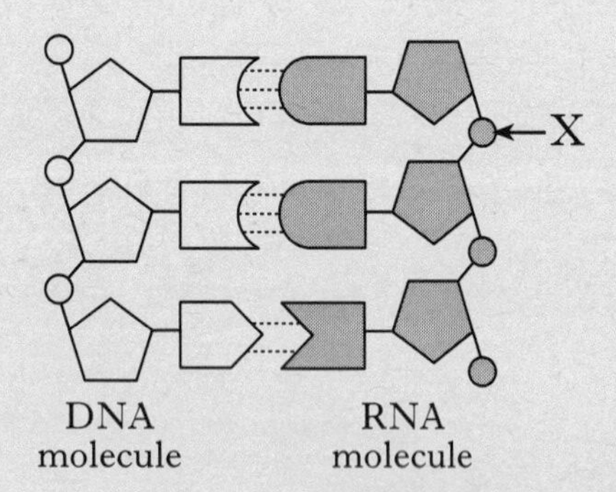

What does component X represent?

A Ribose sugar

B Deoxyribose sugar

C Phosphate

D Ribose phosphate

8. The sequence of triplets on a strand of DNA is shown below.

ATTACACCGTACCAATAG

During translation of mRNA made from the above sequence, how many of the tRNA anticodons will have at least one uracil base?

A 3

B 4

C 5

D 7

9. The function of tRNA in cell metabolism is to

A transport amino acids to be used in synthesis

B carry codons to the ribosomes

C synthesise proteins

D transcribe the DNA code.

10. Which of the following identifies correctly the sequence in which organelles become involved in the production of an enzyme for secretion?

A Nucleus → Ribosomes → Golgi Apparatus → Rough ER

B Ribosomes → Vesicles → Rough ER → Golgi Apparatus

C Nucleus → Rough ER → Vesicles → Ribosomes

D Ribosomes → Rough ER → Golgi Apparatus → Vesicles

11. In a pea plant, the alleles for plant height and petal colour are located on separate chromosomes. The dominant alleles are for tallness and pink petals; the corresponding recessive alleles are for dwarfness and white petals. A heterozygous plant was crossed with a plant recessive for both characteristics. If 320 progeny resulted, what would be the predicted number of tall, white plants?

A 20

B 60

C 80

D 180

12. The relative positions of the genes M, N, O and P on a chromosome were determined by the analysis of percentage recombination. The results are shown in the table.

Genes	*Percentage recombination*
M and O	5
N and O	16
N and P	8
M and P	19

The correct order of genes on the chromosomes is

A O M P N

B O M N P

C M O N P

D M N O P.

13. The base sequence of a short piece of DNA is shown below.

During replication, an inversion mutation occurred on the complementary strand synthesised on this piece of DNA.

Which of the following is the mutated complementary strand?

A T C G A A T G A

B A G C T T A G C

C T C G A A T C G

D T C G A A T G C

14. In a diploid organism with the genotype HhMmNNKK, how many genetically distinct types of gamete would be produced?

A 2

B 4

C 8

D 16

15. Scientists visiting a group of four islands, P, Q, R and S, found similar spiders on each island. They carried out tests to see if the spiders from different islands would interbreed.

The results are summarised in the table below.

(✓ indicates successful interbreeding. ✗ indicates that fertile young were not produced.)

Spiders from	P	Q	R	S
P	✓	✓	✗	✗
Q	✓	✓	✗	✗
R	✗	✗	✓	✗
S	✗	✗	✗	✓

How many species of spider were present on the four islands?

A One

B Two

C Three

D Four

16. In sexual reproduction, which of the following is **not** a source of genetic variation?

A Non-disjunction

B Linkage

C Mutation

D Crossing over

17. Which of the following statements regarding polyploidy is correct?

A It is more common in animals than in plants.

B It is the term used to describe the four haploid cells formed at the end of meiosis.

C It can produce individuals with increased vigour.

D It always results from non-disjunction of chromosomes.

[Turn over

18. In genetic engineering, endonucleases are used to

A join fragments of DNA together

B cut DNA molecules into fragments

C close plasmid rings

D remove cell walls for somatic fusion.

19. Which of the following is a plant response to invasion by a foreign organism?

A Increased production of tannin

B Engulfing of invaders by specialised cells

C Production of antibodies

D Closing of stomata

20. Which of the following adaptations allows a plant to tolerate grazing by herbivores?

A Thick waxy cuticle

B Leaves reduced to spines

C Low meristems

D Thorny stems

Question 21 is at the top of the next column

21. In which of the following do **both** adaptations reduce the rate of water loss from a plant?

A Thin cuticle and rolled leaf

B Rolled leaf and sunken stomata

C Sunken stomata and large surface area

D Thin cuticle and needle-shaped leaves

22. The diagram below shows a transverse section through a plant stem.

In which region would cambium cells be found?

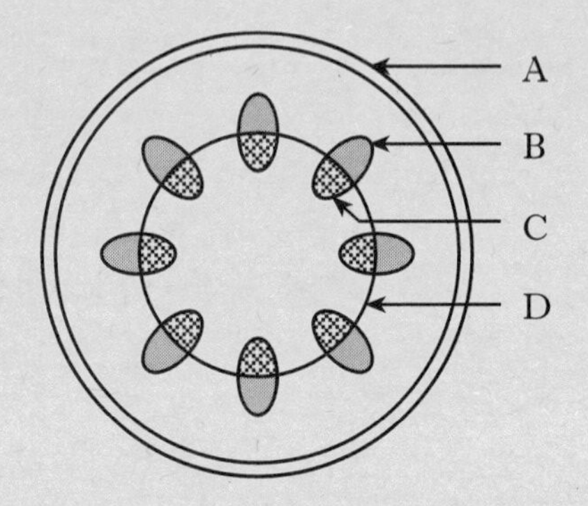

23. The graph below shows the blood glucose concentrations of two women before and after each swallowed 50 g of glucose.

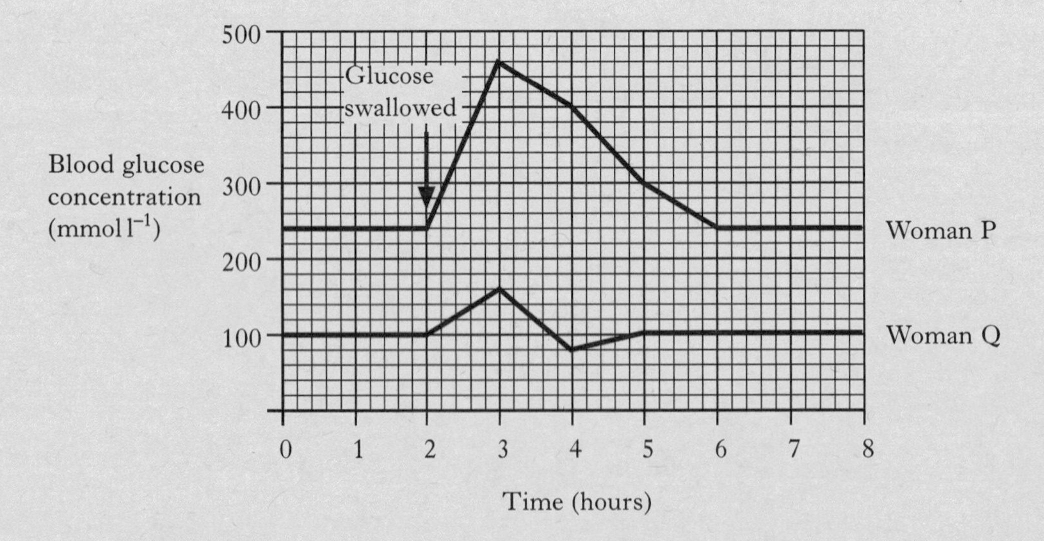

When did the rate of change of blood glucose concentration of the two women differ most?

A Between hours 2 and 3

B Between hours 3 and 4

C Between hours 4 and 5

D Between hours 5 and 6

24. During the germination of barley grains, the plant growth substance GA (Gibberellic Acid) promotes the synthesis of the enzyme α-amylase in the

A aleurone layer

B endosperm

C embryo

D cotyledon.

25. Which of the following statements about the plant growth substances IAA (Indole Acetic Acid) and GA (Gibberellic Acid) is correct?

A An increase in IAA content of a leaf promotes leaf abscission.

B A decrease in IAA content of a leaf promotes leaf abscission.

C An increase in GA content of a leaf promotes leaf abscission.

D A decrease in GA content of a leaf promotes leaf abscission.

26. Which line in the table below identifies correctly the sites of production of the hormones ADH and glucagon?

	ADH	*Glucagon*
A	Pituitary gland	Liver
B	Kidney	Liver
C	Kidney	Pancreas
D	Pituitary gland	Pancreas

27. Which one of the following factors that can limit rabbit population size is density independent?

A Viral disease

B The population of foxes

C The biomass of the grass

D High rainfall

28. Which of the following best defines "population density"?

A The number of individuals present per unit area of a habitat

B The number of individual organisms present in a habitat

C A group of individuals of the same species which make up part of an ecosystem

D The maximum number of individuals which the resources of the environment can support

29. Which of the following does **not** occur during succession from a pioneer community of plants to a climax community?

A Soil fertility increases.

B Larger plants replace smaller plants.

C An increasing intensity of light reaches ground-dwelling plants.

D Each successive community makes the habitat less favourable for itself.

30. Dietary deficiency of vitamin D causes rickets.

This effect is due to

A poor uptake of phosphate into growing bones

B poor calcium absorption from the intestine

C low vitamin D content in the bones

D loss of calcium from the bones.

Candidates are reminded that the answer sheet MUST be returned INSIDE the front cover of this answer book.

[Turn over

DO NOT WRITE IN THIS MARGIN

Marks

SECTION B

All questions in this section should be attempted.

1. (*a*) The diagram below represents cells in the lining of the small intestine of a mammal.

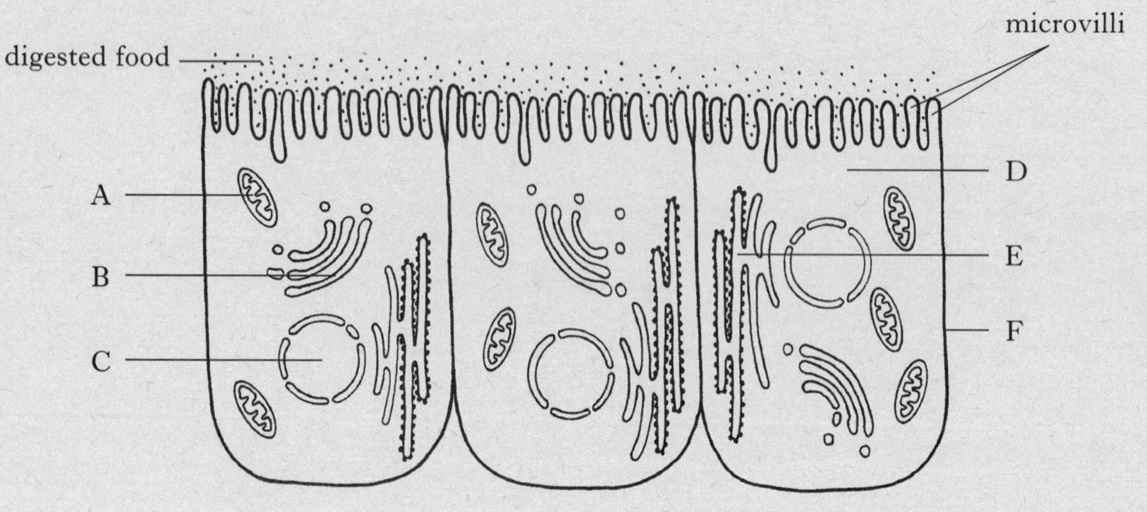

(i) The table below gives information about organelles shown in the diagram.

Complete the table by inserting the appropriate letters, names and functions.

Letter	*Name of organelle*	*Function*
E	Rough endoplasmic reticulum	
		Site of aerobic respiration
B	Golgi apparatus	
		Site of mRNA synthesis

3

(ii) Suggest a reason for the presence of microvilli in this type of cell.

____

____

______________________________ **2**

DO NOT WRITE IN THIS MARGIN

1. (continued)

Marks

(*b*) The diagram below summarises the process of photosynthesis in a chloroplast.

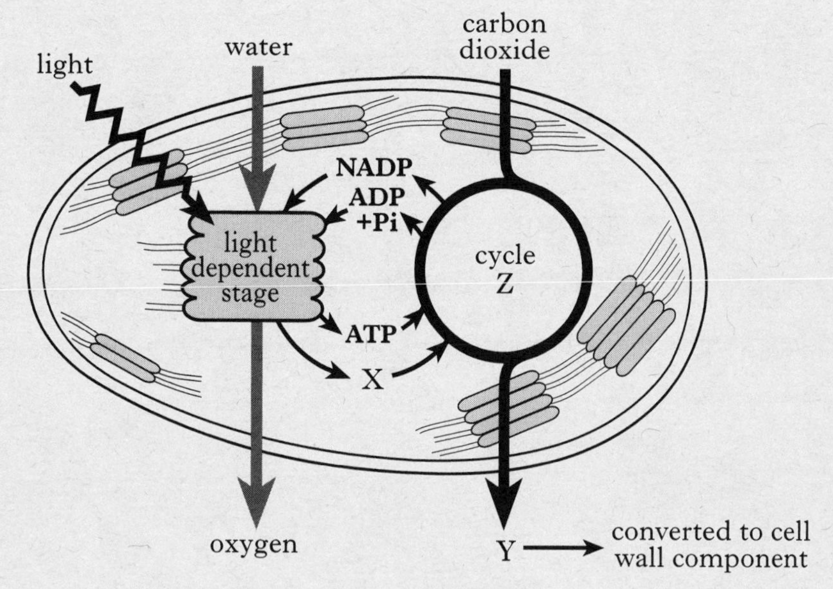

(i) Name molecules X and Y.

X ______________________

Y ______________________ **1**

(ii) State the exact location of the light dependent stage within a chloroplast.

______________________ **1**

(iii) Name cycle Z.

______________________ **1**

(iv) Name the cell wall component referred to in the diagram.

______________________ **1**

[Turn over

Marks

2. An investigation was carried out to compare photosynthesis in oak and nettle leaves.

Six discs were cut from each type of leaf and placed in syringes containing a solution that provided carbon dioxide. A procedure was used to remove air from the leaf discs to make them sink. The apparatus was placed in a darkened room. The discs were then illuminated with a lamp covered with a green filter. Leaf discs which carried out photosynthesis floated.

The positions of the discs one hour later are shown in the diagram below.

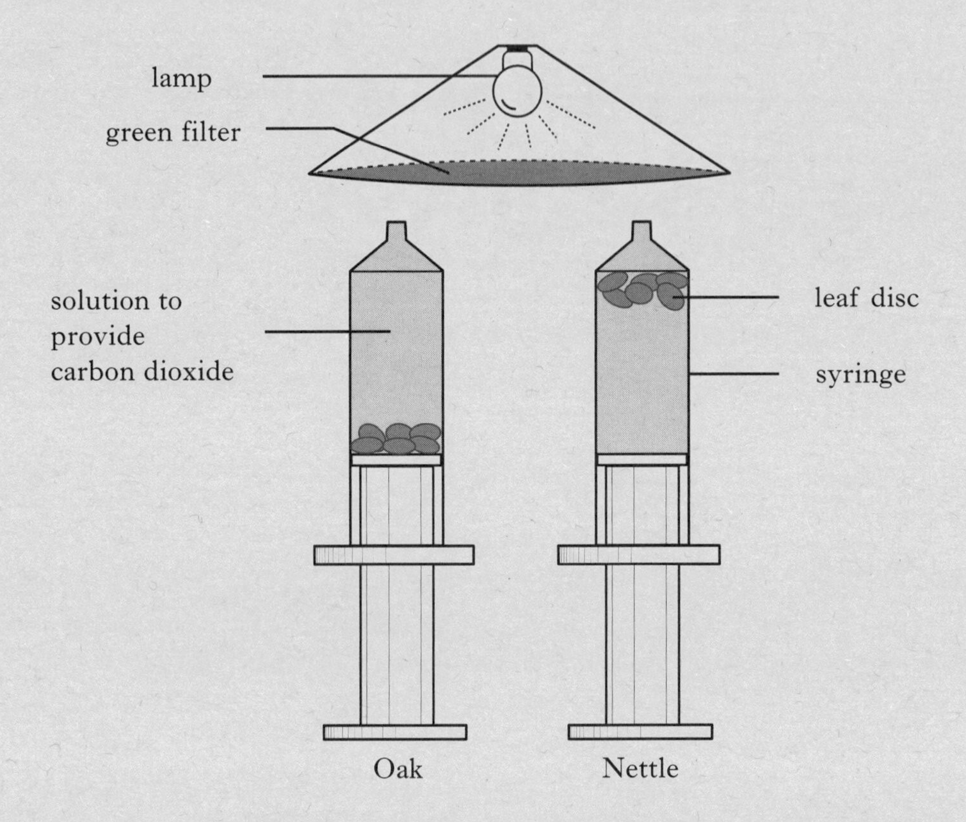

(*a*) Suggest a reason why the investigation was carried out in a darkened room.

____

____ **1**

(*b*) Explain why it was good experimental procedure to use six discs from each plant.

____

____

____ **1**

DO NOT WRITE IN THIS MARGIN

Marks

2. (continued)

(*c*) In setting up the investigation, precautions were taken to ensure that the results obtained would be valid.

Give **one** precaution relating to the preparation of the leaf discs and **one** precaution relating to the solution that provided carbon dioxide.

Leaf discs ____

____ **1**

Solution that provided carbon dioxide ______________________

____ **1**

(*d*) Suggest a reason why the leaf discs which carried out photosynthesis floated.

____ **1**

(*e*) Nettles are shade plants which grow beneath sun plants such as oak trees.

Explain how the results show that nettles are well adapted as shade plants.

____

____

____ **2**

(*f*) What name is given to the light intensity at which the carbon dioxide uptake for photosynthesis is equal to the carbon dioxide output from respiration?

____ **1**

[Turn over

DO NOT WRITE IN THIS MARGIN

2. (continued)

Marks

(*g*) In another investigation, the rate of photosynthesis by nettle leaf discs was measured at different light intensities. The results are shown in the table.

Light intensity (kilolux)	*Rate of photosynthesis by nettle leaf discs* (units)
10	2
20	26
30	58
40	89
50	92
60	92

Plot a line graph to show the rate of photosynthesis by nettle leaf discs at different light intensities. Use appropriate scales to fill most of the graph paper.
(Additional graph paper, if required, can be found on page 32.)

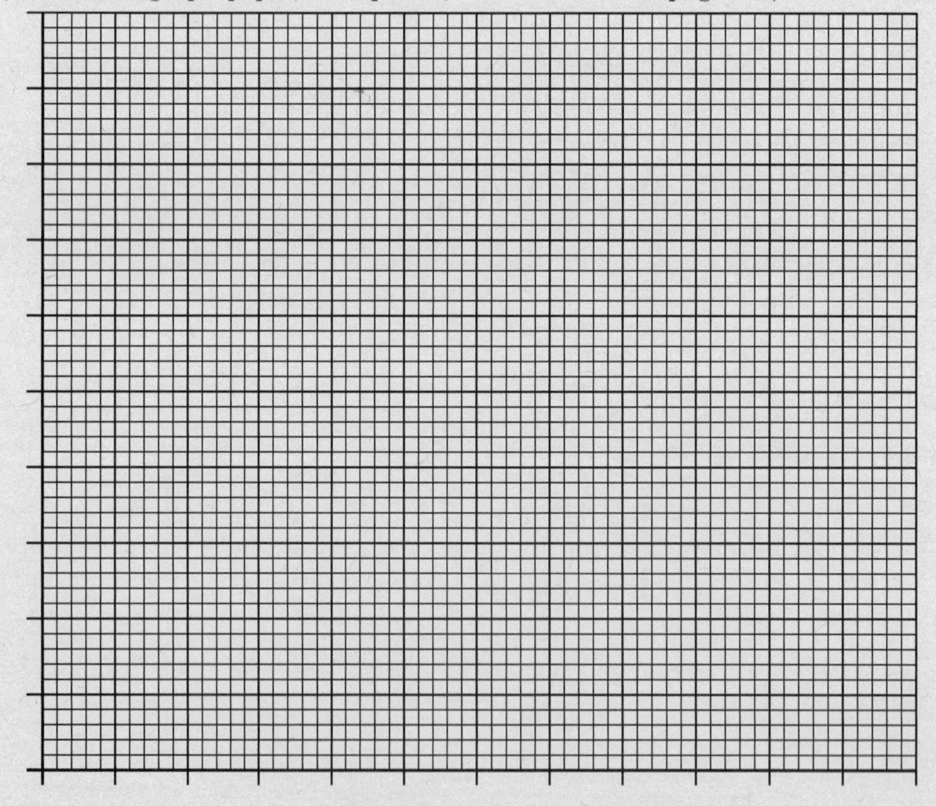

2

(*h*) From the table, predict how the rate of photosynthesis at a light intensity of 50 kilolux could be affected by an increase in carbon dioxide concentration. Justify your answer.

Effect on the rate of photosynthesis ______________________________

Justification ______________________________

______________________________ **1**

DO NOT WRITE IN THIS MARGIN

Marks

3. The stages shown below take place when a human cell is invaded by an influenza virus.

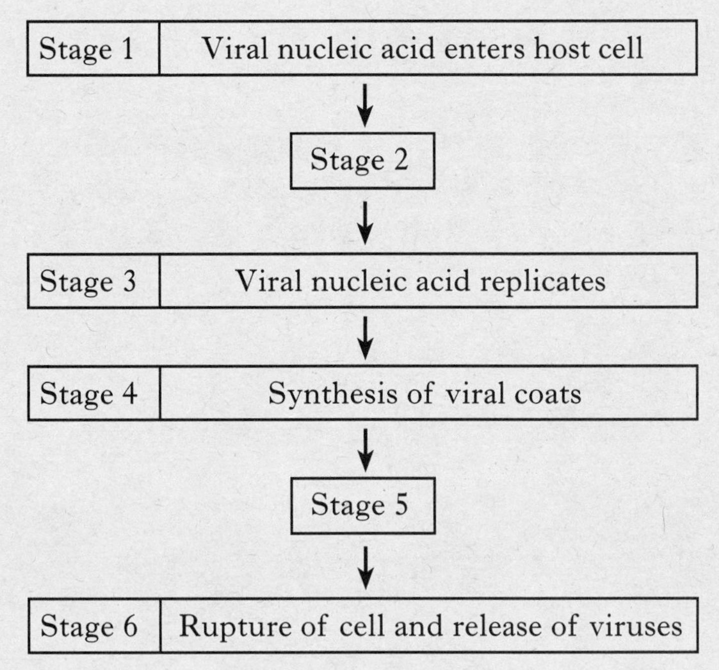
Stage 1
Viral nucleic acid enters host cell
Stage 2
Stage 3
Viral nucleic acid replicates
Stage 4
Synthesis of viral coats
Stage 5
Stage 6
Rupture of cell and release of viruses

(*a*) Describe the processes that occur during Stages 2 and 5.

Stage 2 ______________________________

______________________________ **1**

Stage 5 ______________________________

______________________________ **1**

(*b*) Name the cell organelle at which the viral coats are synthesised during Stage 4.

______________________________ **1**

(*c*) During a viral infection, a type of white blood cell is stimulated to make antibodies which inactivate the viruses.

(i) Name this type of white blood cell.

______________________________ **1**

(ii) What feature of viruses stimulates these cells to make antibodies?

______________________________ **1**

(iii) New strains of influenza virus appear regularly. Suggest why antibodies produced against one strain of virus are not effective against another strain.

____

______________________________ **1**

Marks

4. An outline of the process of respiration is shown in the diagram below.

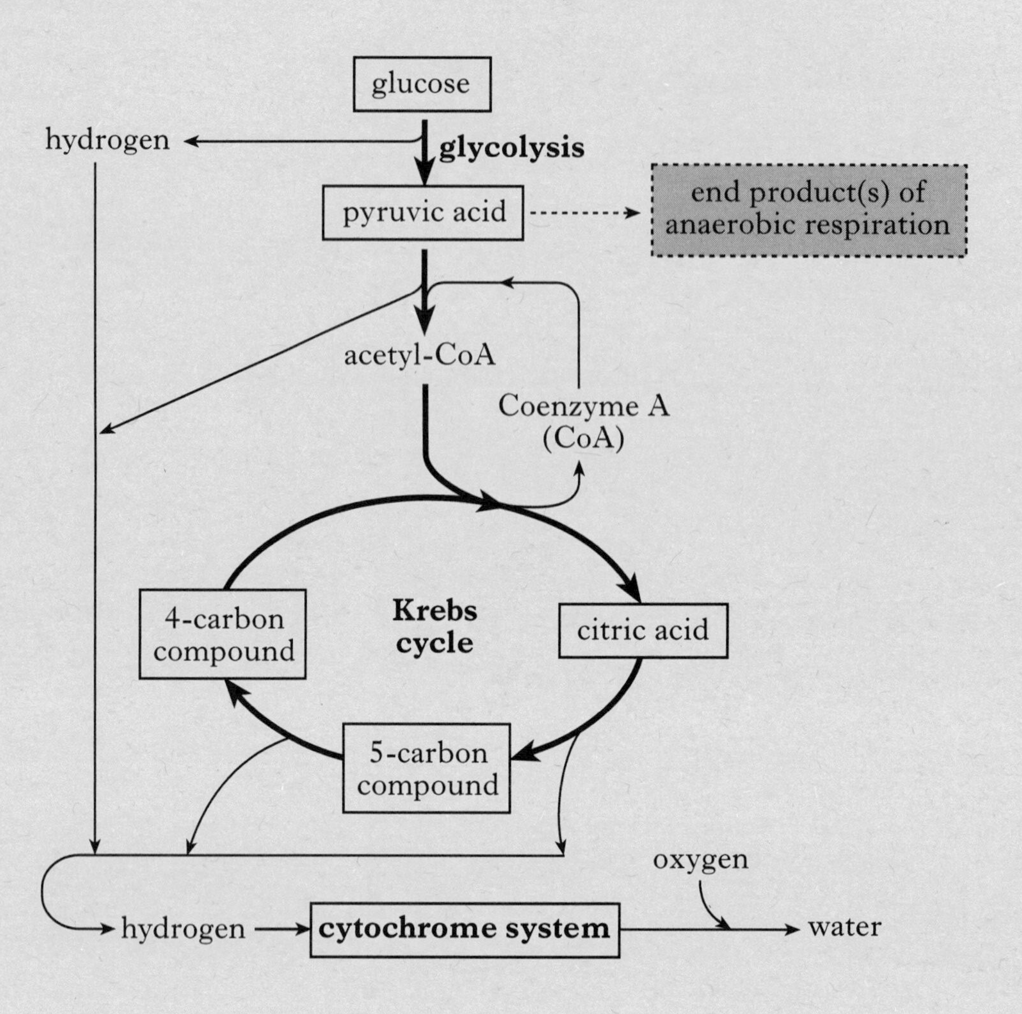

(*a*) Apart from glucose and enzymes, what chemical substance is essential for glycolysis to occur?

______________________________ **1**

(*b*) Name the end-product(s) of anaerobic respiration in an animal cell and a plant cell.

(i) Animal cell ____________________ **1**

(ii) Plant cell ____________________ **1**

(*c*) Name the carrier that transfers hydrogen to the cytochrome system.

______________________________ **1**

DO NOT WRITE IN THIS MARGIN

Marks

4. (continued)

(*d*) Explain why the cytochrome system cannot function in anaerobic conditions.

____

____

____ **1**

(*e*) The energy content of glucose is 2900 kJ mol^{-1} and during aerobic respiration 1178 kJ mol^{-1} of this energy is stored in ATP.

Calculate the percentage of the energy content of glucose that is stored in ATP.

Space for calculation

__________% **1**

(*f*) Which stage of respiration releases **most** energy for use by the cell?

____ **1**

[Turn over

DO NOT WRITE IN THIS MARGIN

Marks

5. The diagram below represents a stage of meiosis in a cell from a female fruit fly, *Drosophila*.

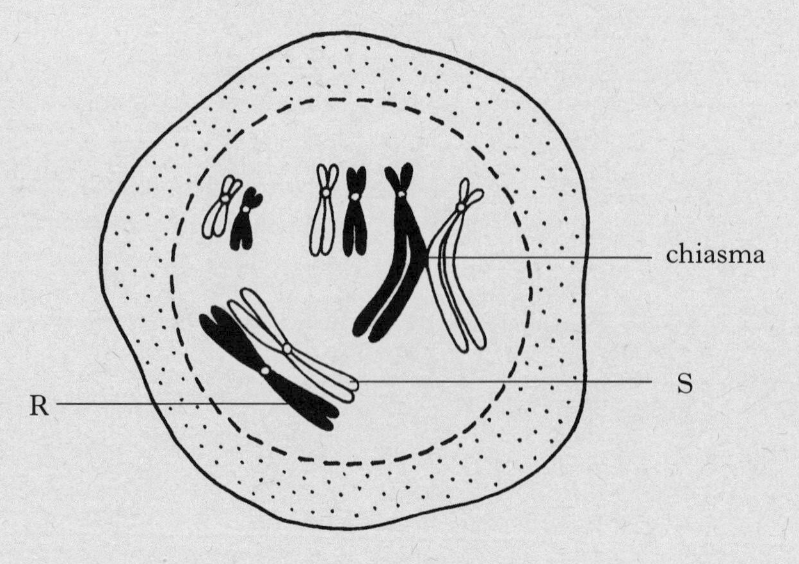

(*a*) Name the tissue from which this cell was taken.

____ **1**

(*b*) What is the haploid number of this species?

____________________ **1**

(*c*) Chromosomes R and S are homologous. Apart from their appearance, state **one** similarity between homologous chromosomes.

____

____ **1**

(*d*) Explain the importance of chiasmata formation.

____

____ **1**

DO NOT WRITE IN THIS MARGIN

Marks

6. In humans, the allele for red-green colour deficiency (b) is sex-linked and recessive to the normal allele (B).

The family tree diagram below shows how the condition was inherited.

□ Male without the condition

■ Male with the condition

○ Female without the condition

● Female with the condition

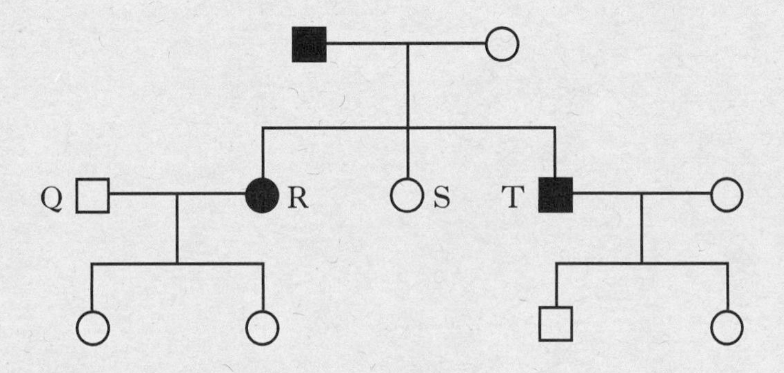

(*a*) Give the genotypes of individuals S and T.

(i) S ____________________________ **1**

(ii) T ____________________________ **1**

(*b*) If individuals Q and R have a son, what is the chance that he will inherit the condition?

Space for calculation

Chance ____________________ **1**

(*c*) Explain why individual R has the condition although her mother was unaffected.

____

____ **1**

[Turn over

Marks

7. Hawaii is a group of islands isolated in the Pacific Ocean.

Different species of Honeycreeper birds live on these islands.

The heads of four species of Honeycreeper are shown below.

(*a*) (i) Explain how the information given about Honeycreeper species supports the statement that they occupy different niches.

____

____ **1**

(ii) What further information would be needed about the four species of Honeycreeper to conclude that they had evolved by adaptive radiation?

____

____ **1**

(*b*) The Honeycreeper species have evolved in geographical isolation.

Name **one** other type of isolating barrier involved in the evolution of new species.

____ **1**

DO NOT WRITE IN THIS MARGIN

Marks

8. The marine worm *Sabella* lives in a tube made out of sand grains from which it projects a fan of tentacles for feeding.

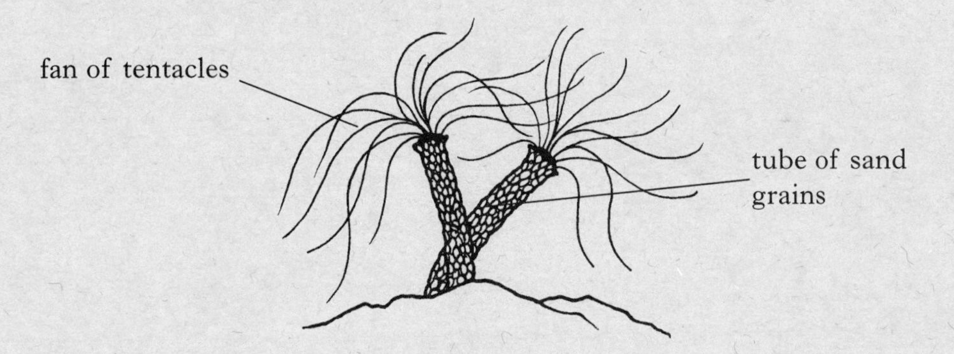

(*a*) If the worm is disturbed, the fan is immediately withdrawn into the tube. The fan re-emerges a few minutes later.

(i) Name the type of behaviour illustrated by the withdrawal response.

____ **1**

(ii) What is the advantage to the worm of withdrawing its tentacles in response to a disturbance?

____

____ **1**

(*b*) If a harmless stimulus occurs repeatedly, the withdrawal response eventually ceases.

(i) Name the type of behaviour illustrated by this modified response.

____ **1**

(ii) What is the advantage to the worm of this modified response?

____

____ **1**

[Turn over

DO NOT WRITE IN THIS MARGIN

9. Limpets (*Patella*) feed by grazing on algae growing on rocks at the seashore.

Marks

limpet

Limpet shell

height

length

Graph 1 below shows the effects of limpet population density on the average shell length and total biomass.

Graph 1

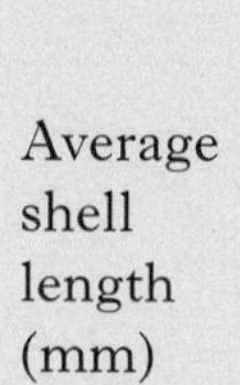

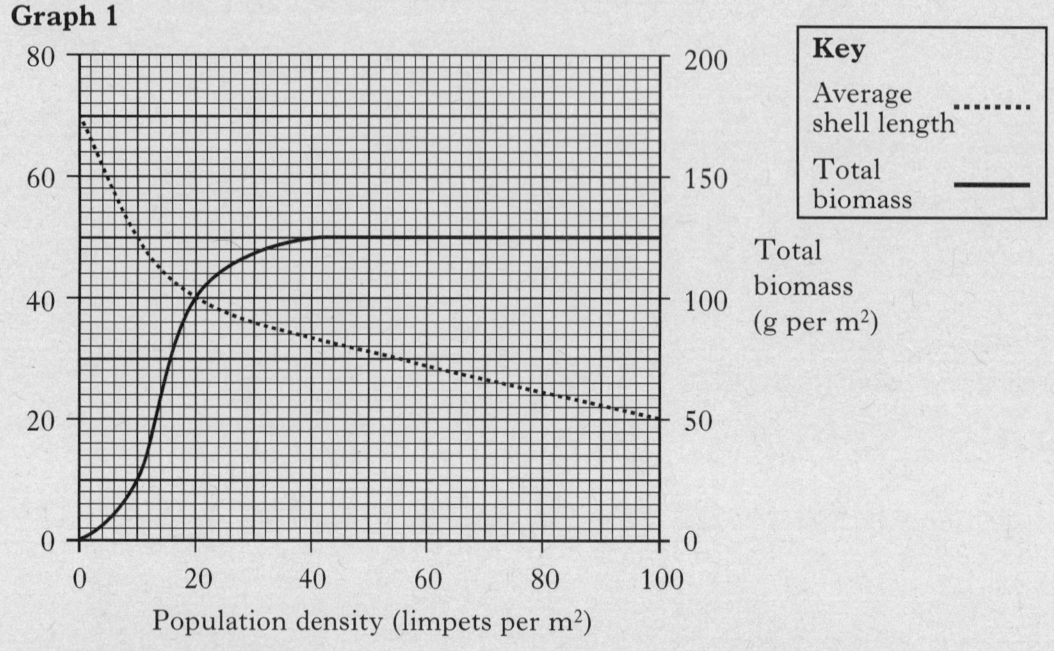

(*a*) What is the total biomass at a population density of 10 limpets per m^2?

____________________ g per m^2 **1**

(*b*) Identify the population density range (limpets per m^2) in which the total biomass increases most rapidly.

Tick the correct box.

0–10 ☐ 10–20 ☐ 20–30 ☐ 30–40 ☐ 40–50 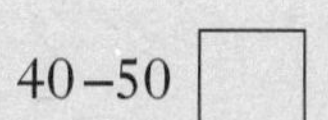 **1**

(*c*) Calculate the average mass of one limpet when the population density is 20 per m^2.

Space for calculation

Average mass ____________ g **1**

(*d*) Use values from Graph 1 to describe the effect of increasing population density on the total biomass of limpets.

____

____

____ **2**

DO NOT WRITE IN THIS MARGIN

9. (continued)

Marks

(*e*) Explain how intraspecific competition causes the trend in average shell length shown in Graph 1.

____

____ **1**

(*f*) The table below shows information about limpets on shore A which is sheltered and on shore B which is exposed to strong wave action.

Graph 2 below shows the effect of wave action on limpet shell index.

$$\text{Limpet shell index} = \frac{\text{shell height}}{\text{shell length}}$$

Shore A (sheltered)		*Shore B (exposed)*	
Shell height (mm)	*Shell length* (mm)	*Shell height* (mm)	*Shell length* (mm)
16	52	9	21
19	54	11	26
20	55	14	31
21	56	16	34
22	57	17	35
23	58	17	36
26	60	–	–
Average = 21	Average =	Average = 14	Average =

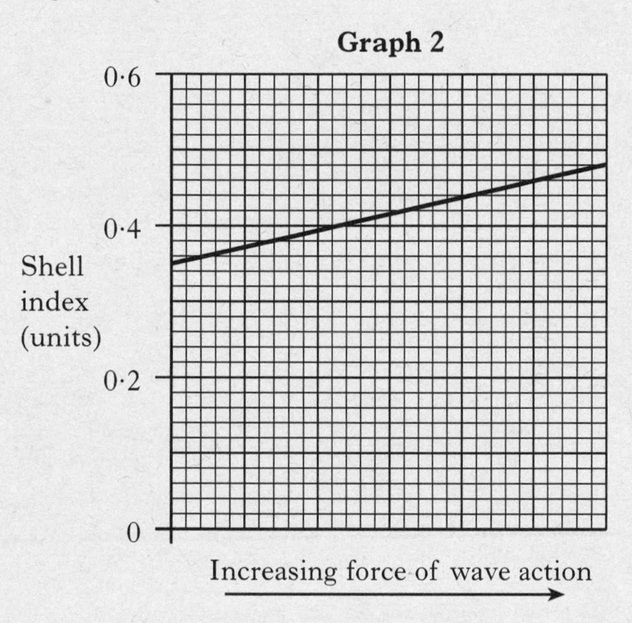

(i) **Complete the table** by calculating the average shell length of limpets on both shores.

Space for calculation

1

(ii) Express as the **simplest whole number ratio** the average shell height for shore A and shore B.

Space for calculation

Ratio _______ : _______ **1**

(iii) A limpet shell collected on one of the shores had a length of 43 mm and a height of 20 mm. Use Graph 2 to identify which shore it came from and justify your choice.

Tick (✓) the correct box Shore A ☐ Shore B ☐

Justification ____

____

____ **1**

Marks

10. (*a*) The grid below shows adaptations of bony fish for osmoregulation.

A	few, small glomeruli	**B**	active secretion of salts by gills	**C**	high filtration rate in kidney
D	active uptake of salts by gills	**E**	low filtration rate in kidney	**F**	many, large glomeruli

Use letters from the grid to answer the following questions.

(i) Which **three** adaptations would be found in freshwater fish?

Letters _______, _______ and _______ . **1**

(ii) Which **two** adaptations would result in the production of a small volume of urine?

Letters _______ and _______ . **1**

(*b*) The table shows some adaptations of a desert mammal which help to conserve water.

For each adaptation, tick (✓) the correct box to show whether it is behavioural **or** physiological.

Adaptation	*Behavioural*	*Physiological*
High level of blood ADH		
Lives in underground burrow		
Nocturnal foraging		
Absence of sweating		

2

DO NOT WRITE IN THIS MARGIN

Marks

11. (*a*) The diagram below shows a section through part of a root.

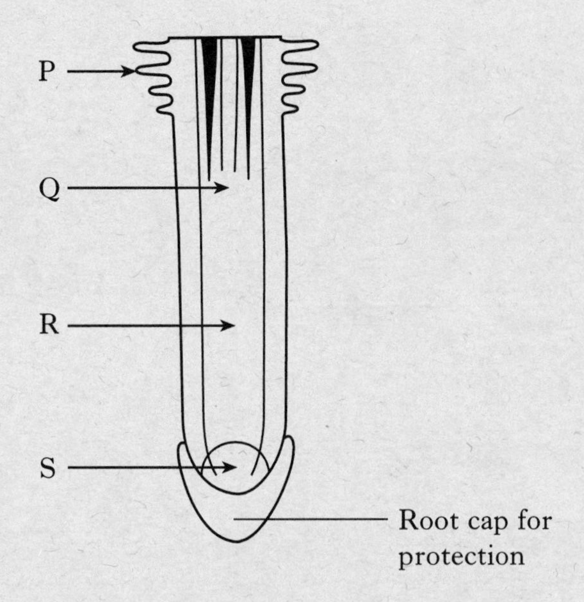

(i) Which letter shows the position of a meristem?

Letter ______________ **1**

(ii) Name a cell process responsible for increase in length of a root.

______________________________ **1**

(*b*) The diagram below shows the growth pattern of a locust.

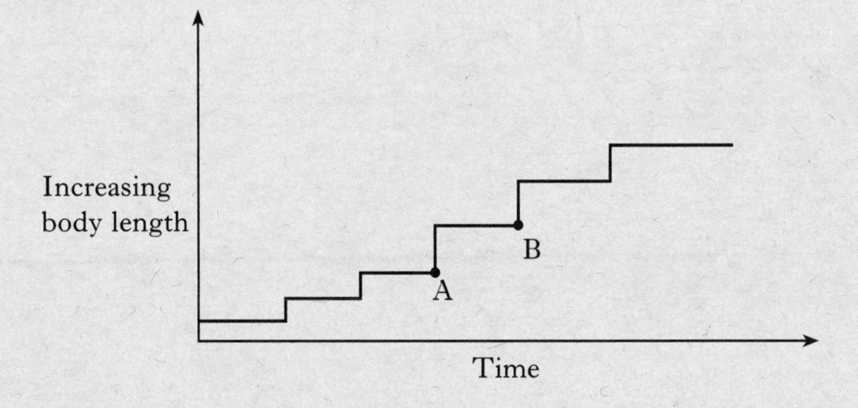

Explain the reason for the shape of the growth pattern between A and B.

____

____

______________________________ **2**

[Turn over

Marks

12. The diagram below shows the apparatus used to investigate the growth of oat seedlings in water culture solutions. Each solution lacks one element required for normal growth.

The containers were painted black to prevent algal growth.

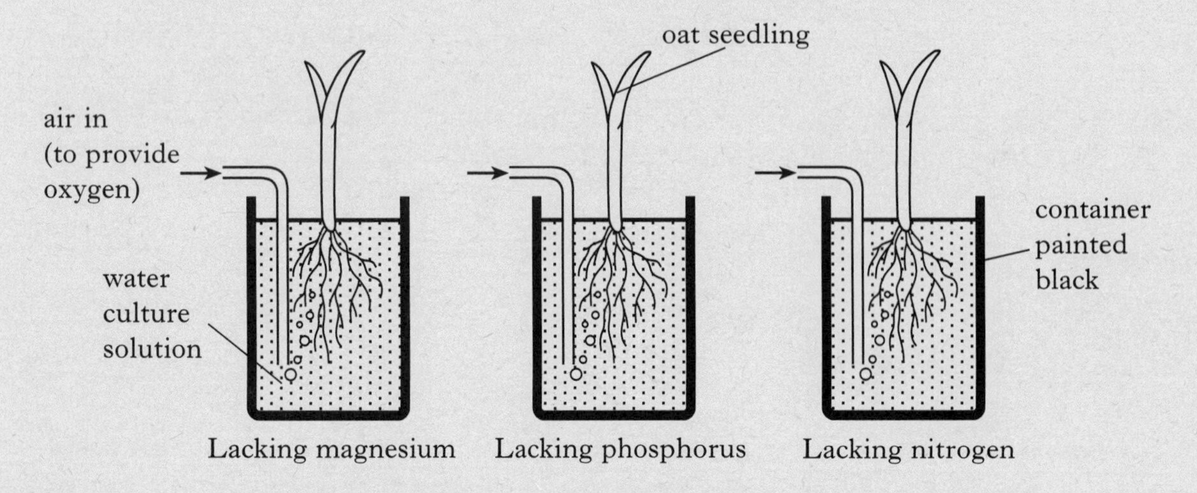

(*a*) Describe a suitable control for this experiment.

1

(*b*) Suggest a reason why algal growth should be prevented in the culture solutions during the investigation.

1

(*c*) The table below shows the elements investigated and symptoms of their deficiency.

Place ticks (✓) in the correct boxes to match each element with the symptoms of its deficiency.

Element	*Symptoms of deficiency*	
	Leaf bases red	*Chlorotic leaves*
Magnesium		
Phosphorus		
Nitrogen		

2

DO NOT WRITE IN THIS MARGIN

Marks

12. (continued)

(*d*) Name a magnesium containing molecule found in oat seedlings.

______________________________ **1**

(*e*) Explain why the uptake of elements by oat seedling roots is dependent on the availability of oxygen.

____

____

______________________________ **2**

[Turn over

DO NOT WRITE IN THIS MARGIN

Marks

13. The production of thyroxine in mammals is controlled by the hormone TSH. Thyroxine controls metabolic rate in body cells and has a negative feedback effect on gland X.

The diagram below shows the relationship between TSH and thyroxine production.

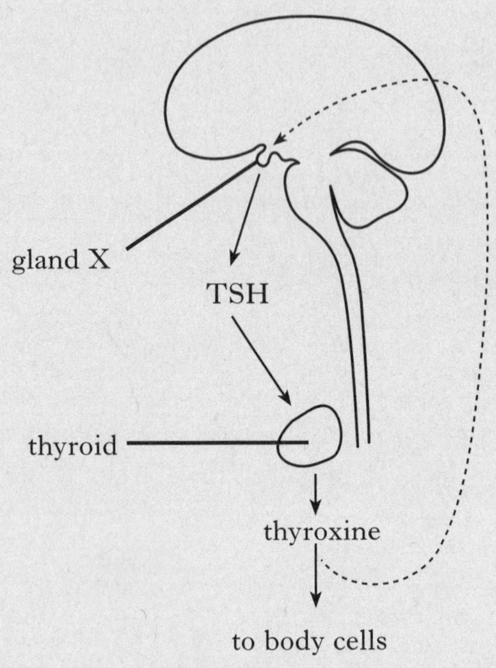

(*a*) Name gland X.

____ **1**

(*b*) In an investigation into the effect of thyroxine, groups of rats of similar mass were treated as follows.

Group A were fed a normal diet.

Group B were fed a normal diet plus thyroxine.

Group C were fed a normal diet plus an inhibitor of thyroxine production.

The table below shows the average hourly oxygen consumption in cm^3 per gram of body mass in rats from each group.

Group	*Average hourly oxygen consumption* (cm^3g^{-1})
A	1·6
B	2·8
C	1·2

(i) Explain how the results in the table support the statement that an increase in metabolic rate leads to an increase in oxygen consumption.

____

____

____ **2**

DO NOT WRITE IN THIS MARGIN

Marks

13. (*b*) (continued)

(ii) What evidence suggests that rats fed a normal diet make thyroxine?

____

____ 1

(iii) How would the level of TSH production in group A compare with group C?

____

____ 1

(iv) Calculate the percentage decrease in oxygen consumption which results from feeding the thyroxine inhibitor to rats.

Space for calculation

__________ % decrease 1

(v) The table below relates to aspects of the appearance and behaviour of rats in groups B and C.

Group	*Appearance of ears and feet*	*Behaviour*
B	Pink	Lie stretched out
C	Pale	Lie curled up with feet tucked in

Complete the following sentences by underlining **one** of the alternatives in each pair.

1 Compared with rats in group B, the rats in group C have a {lower / higher} metabolic rate and show {dilation / constriction} of skin blood vessels. 1

2 The behaviour of rats in group C allows them to {lose / conserve} body heat. 1

[Turn over

DO NO WRITE THIS MARGI

SECTION C

Marks

Both questions in this section should be attempted.

Note that each question contains a choice.

Questions 1 and 2 should be attempted on the blank pages which follow.

Supplementary sheets, if required, may be obtained from the invigilator.

Labelled diagrams may be used where appropriate.

1. Answer **either** A **or** B.

A. Give an account of gene mutation under the following headings:

(i) the occurrence of mutant alleles and the effect of mutagenic agents; **3**

(ii) types of gene mutation and how they alter amino acid sequences. **7**

(10)

OR

B. Give an account of water movement through plants under the following headings:

(i) the transpiration stream; **8**

(ii) importance of the transpiration stream. **2**

(10)

In question 2, ONE mark is available for coherence and ONE mark is available for relevance.

2. Answer **either** A **or** B.

A. Give an account of the mechanisms and importance of temperature regulation in endotherms. **(10)**

OR

B. Give an account of the effect of light on shoot growth and development, and on the timing of flowering in plants and breeding in animals. **(10)**

[*END OF QUESTION PAPER*]

DO NOT WRITE IN THIS MARGIN

SPACE FOR ANSWERS

SPACE FOR ANSWERS

DO NOT WRITE IN THIS MARGIN

SPACE FOR ANSWERS

SPACE FOR ANSWERS

ADDITIONAL GRAPH PAPER FOR QUESTION 2(*g*)

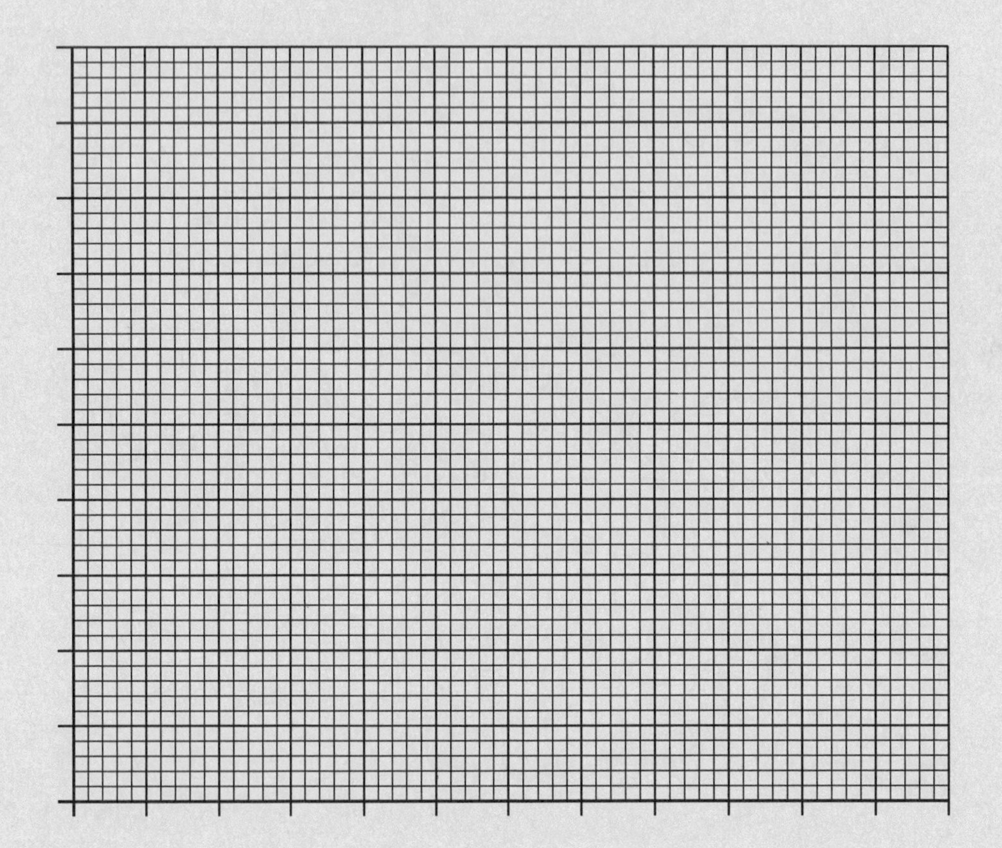

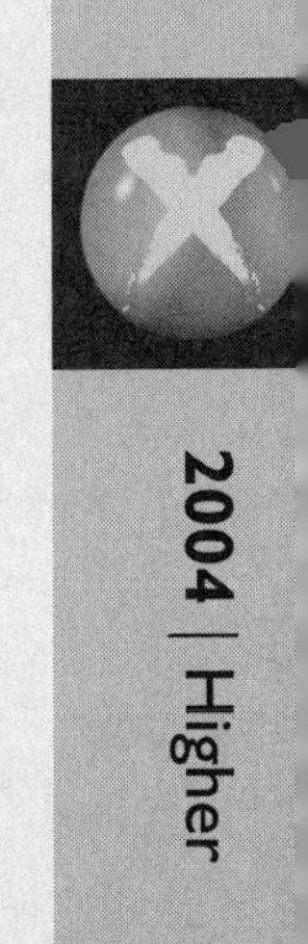

[BLANK PAGE]

FOR OFFICIAL USE

Total for Sections B and C

X007/301

NATIONAL QUALIFICATIONS 2004

WEDNESDAY, 19 MAY
1.00 PM – 3.30 PM

BIOLOGY
HIGHER

Fill in these boxes and read what is printed below.

Full name of centre

Town

Forename(s)

Surname

Date of birth
Day Month Year

Scottish candidate number

Number of seat

SECTION A—Questions 1–30 (30 marks)

Instructions for completion of Section A are given on page two.

SECTIONS B AND C (100 marks)

1 (a) All questions should be attempted.
  (b) It should be noted that in **Section C** questions 1 and 2 each contain a choice.

2 The questions may be answered in any order but all answers are to be written in the spaces provided in this answer book, and must be written clearly and legibly in ink.

3 Additional space for answers and rough work will be found at the end of the book. If further space is required, supplementary sheets may be obtained from the invigilator and should be inserted inside the **front** cover of this book.

4 The numbers of questions must be clearly inserted with any answers written in the additional space.

5 Rough work, if any should be necessary, should be written in this book and then scored through when the fair copy has been written.

6 Before leaving the examination room you must give this book to the invigilator. If you do not, you may lose all the marks for this paper.

SCOTTISH QUALIFICATIONS AUTHORITY

LIB X007/301 6/13470

SECTION A

Read carefully

1 Check that the answer sheet provided is for Biology Higher (Section A).

2 Fill in the details required on the answer sheet.

3 In this section a question is answered by indicating the choice A, B, C or D by a stroke made in **ink** in the appropriate place in the answer sheet—see the sample question below.

4 For each question there is only **one** correct answer.

5 Rough working, if required, should be done only on this question paper—or on the rough working sheet provided—**not** on the answer sheet.

6 At the end of the examination the answer sheet for Section A **must** be placed inside the front cover of this answer book.

Sample Question

The apparatus used to determine the energy stored in a foodstuff is a

A respirometer

B calorimeter

C klinostat

D gas burette.

The correct answer is **B**—calorimeter. A **heavy** vertical line should be drawn joining the two dots in the appropriate box in the column headed **B** as shown in the example on the answer sheet.

If, after you have recorded your answer, you decide that you have made an error and wish to make a change, you should cancel the original answer and put a vertical stroke in the box you now consider to be correct. Thus, if you want to change an answer D to an answer B, your answer sheet would look like this:

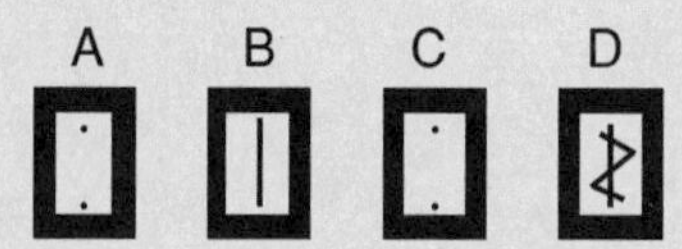

If you want to change back to an answer which has already been scored out, you should enter a tick (✓) to the **right** of the box of your choice, thus:

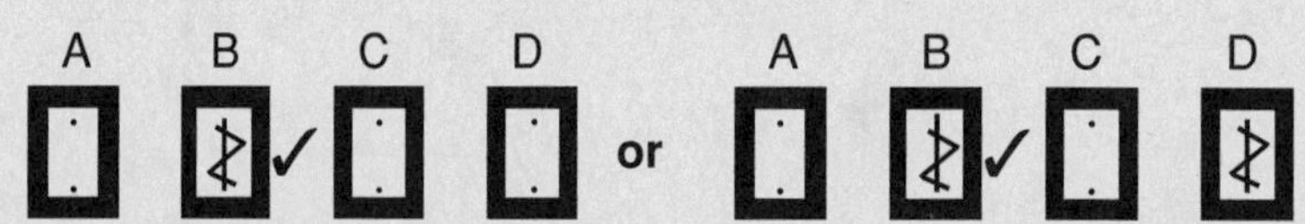

SECTION A

All questions in this section should be attempted.

Answers should be given on the separate answer sheet provided.

1. Which of the following processes requires infolding of the cell membrane?

A Diffusion

B Phagocytosis

C Active transport

D Osmosis

2. The diagram shows the fate of sunlight landing on a leaf.

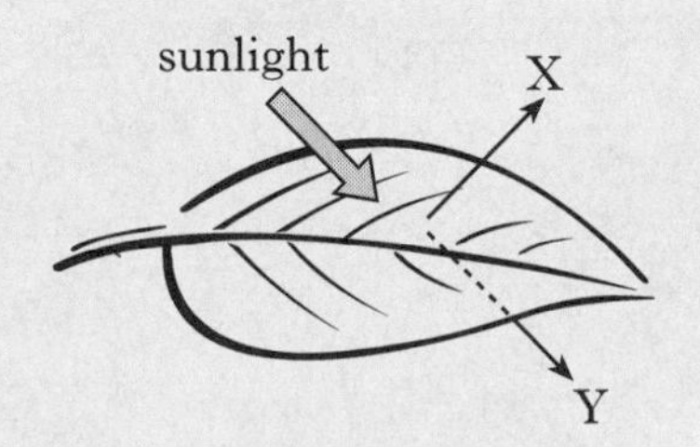

Which line in the table below identifies correctly the fate of sunlight represented by X and Y?

	X	*Y*
A	transmission	reflection
B	absorption	transmission
C	reflection	transmission
D	reflection	absorption

3. Which of the following colours of light are mainly absorbed by chlorophyll a?

A Orange and violet

B Blue and red

C Blue and green

D Green and orange

4. The graph shows the effect of temperature on the rate of reactions in the light dependent stage in photosynthesis.

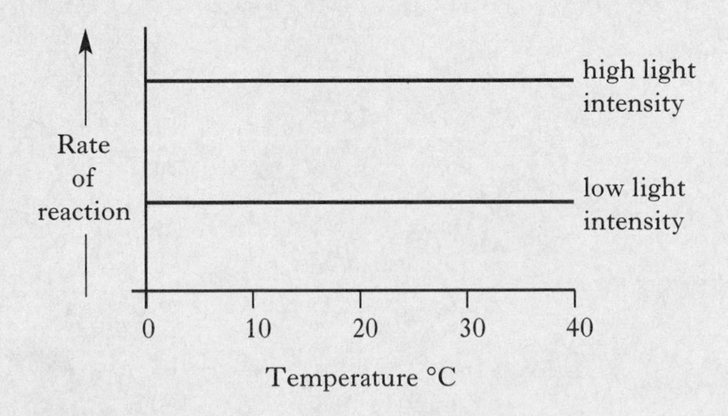

From the graph, it may be deduced that

A enzymes are not involved in controlling these reactions

B enzymes act most effectively at high intensities of light

C at the high intensity of light, carbon dioxide is the limiting factor

D the rate of the reaction increases with increase in temperature.

5. The graph shows the effect of increasing light intensity on the rate of photosynthesis.

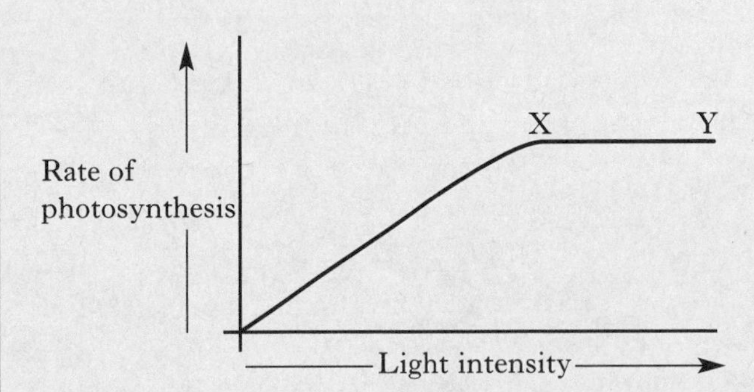

Two environmental factors which could limit the rate of photosynthesis between points X and Y are

A light intensity and oxygen concentration

B temperature and light intensity

C temperature and carbon dioxide concentration

D carbon dioxide concentration and light intensity.

 [Turn over

6. In respiration, the sequence of reactions resulting in the conversion of glucose to pyruvic acid is called

A the Krebs cycle

B the citric acid cycle

C glycolysis

D the cytochrome chain.

7. Which line in the table describes correctly both aerobic respiration and anaerobic respiration in human muscle tissue?

	Aerobic Respiration	*Anaerobic Respiration*
A	There is a net gain of ATP	Carbon dioxide is not produced
B	There is a net gain of ATP	Oxygen is required
C	Carbon dioxide is produced	There is a net loss of ATP
D	Lactic acid is formed	Ethanol is formed

8. Cyanogenesis in *Trifolium repens* is a defence mechanism against

A water loss

B fungal infection

C bacterial invasion

D grazing.

9. A sex-linked gene carried on the X-chromosome of a man will be transmitted to

A 50% of his male children

B 50% of his female children

C 100% of his male children

D 100% of his female children.

10. The inheritance of eye colour in *Drosophila* is sex-linked and the allele for red eyes (R) is dominant to the allele for white eyes (r).

The progeny of a cross were all red-eyed females and white-eyed males.

What were the genotypes of their parents?

A X^rX^r X^RY

B X^RX^r X^RY

C X^RX^r X^rY

D X^RX^R X^rY

11. Black coat colour in cocker spaniels is determined by a dominant gene (B) and red coat colour by its recessive allele (b). Uniform coat colour is determined by a dominant gene (F) and spotted coat colour by its recessive allele (f).

A male with a uniform black coat was mated to a female with a uniform red coat. A litter of six pups was produced, two of which had uniform black coat colour, two had uniform red coat colour, one had spotted black coat colour and one had spotted red coat colour.

The genotypes of the parents were

A BBFf × bbFf

B BbFf × bbFF

C BbFf × BbFf

D BbFf × bbFf.

12. A tall plant with purple petals was crossed with a dwarf plant with white petals. The F_1 generation were all tall plants with purple petals.

The F_1 generation was self pollinated and produced 1600 plants.

Which line in the table identifies correctly the most likely phenotypic ratio in the F_2 generation?

	Tall purple	*Tall white*	*Dwarf purple*	*Dwarf white*
A	870	325	305	100
B	870	0	0	730
C	400	400	400	400
D	530	260	270	540

13. The table below shows the percentage recombination frequencies for four genes present on the same chromosome.

Gene pair	*% recombination frequency*
P and Q	33
R and Q	40
R and S	32
P and R	7
Q and S	8

Which of the following represents the correct order of genes on the chromosome?

A Q P S R

B P Q S R

C Q S P R

D P Q R S

14. Which of the following describes the term non-disjunction?

A The failure of chromosomes to separate at meiosis.

B The independent assortment of chromosomes at meiosis.

C The exchange of genetic information at chiasmata.

D An error in the replication of DNA before cell division.

15. Which of the following is true of polyploid plants?

A They have reduced vigour and the diploid chromosome number.

B They have increased vigour and the diploid chromosome number.

C They have reduced vigour and sets of chromosomes greater than the diploid chromosome number.

D They have increased vigour and sets of chromosomes greater than the diploid chromosome number.

16. Somatic fusion is a technique which is used to

A fuse cells from different species of animal

B fuse cells from different species of plant

C transfer genetic information into a bacterium

D alter the genes carried on a plasmid.

[Turn over

17. The graph shows the carbon dioxide gain or loss in a shade plant and in a sun plant during part of a day in summer.

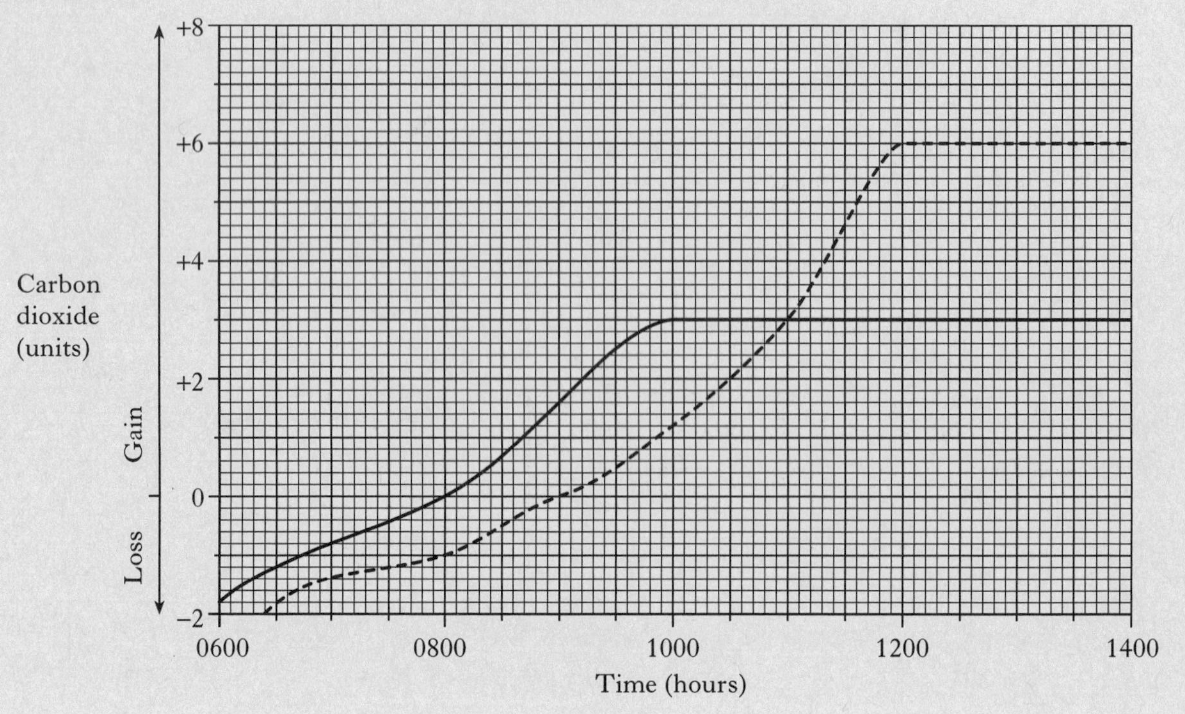

At what time does the shade plant reach compensation point?

A 0800 hours

B 0900 hours

C 1000 hours

D 1200 hours

---

18. The table shows water gain and loss in a plant on two consecutive days.

	Water gain (cm^3)	*Water loss* (cm^3)
First day	100	120
Second day	95	90

Conditions on the second day may have differed from conditions on the first day in some of the following ways.

1 Higher temperature

2 Lower windspeed

3 Lower humidity

4 Lower temperature

Which two conditions could account for the differences in water gain and loss from the first day to the second day?

A 1 and 2

B 1 and 3

C 2 and 4

D 3 and 4

19. Grass can survive despite being grazed by herbivores such as sheep and cattle. It is able to tolerate grazing because it

A is a wind-pollinated plant

B grows constantly throughout the year

C possesses poisons which protect it from being eaten entirely

D has very low growing points which send up new leaves when older ones are eaten.

20. When the intensity of grazing by herbivores increases in a grassland ecosystem, diversity of plant species may increase as a result.

Which statement explains this observation?

A Few herbivores are able to graze on every plant species present.

B Grazing stimulates growth in some plant species.

C Vigorous plant species are grazed so weaker competitors can also thrive.

D Plant species with defences against grazing are selected.

21. Which of the following describes an advantage of habituation to an animal?

A The animal becomes very good at an action which is performed repeatedly.

B An animal shows the same behaviour patterns as all those of the same species.

C A particular response is learned very quickly.

D Energy is not wasted in responding to harmless stimuli.

22. Which of the following examples of bird behaviour would result in reduced interspecific competition?

A Great Tits with the widest stripe on their breast feed first when food is scarce.

B Sooty Terns feed on larger fish than other species of tern which live in the same area.

C Pelicans searching for food form a large circle round a shoal of fish, then dip their beaks into the water simultaneously.

D Predatory gulls have difficulty picking out an individual puffin from a large flock.

23. The table shows the relative percentages by mass of the major chemical groups in a sample of human tissue.

The remaining percentage is made up of water.

Chemical group	%
Carbohydrate	5
Protein	18
Lipid	10
Other organic material	2
Inorganic material	1

What mass of water is present in a 250 g sample of this tissue?

A 64 g

B 36 g

C 90 g

D 160 g

[Turn over

24. The diagram below shows the human body's responses to temperature change.

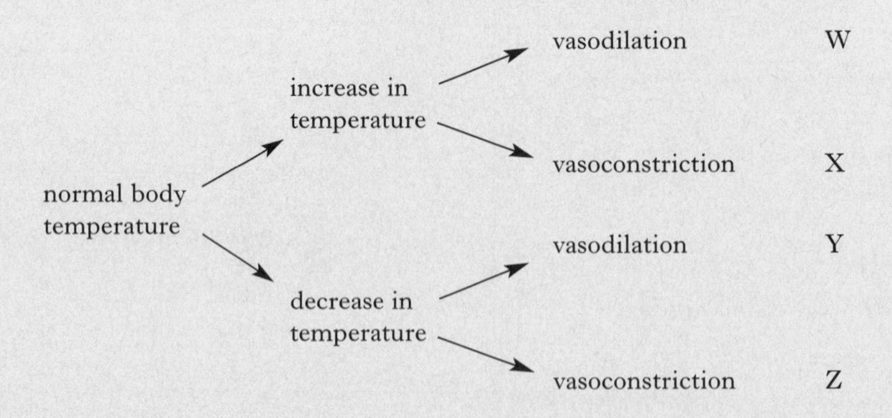

Which letters indicate negative feedback control of body temperature?

A W and Y

B W and Z

C X and Y

D X and Z

25. Muscle cells differ from nerve cells because

A they contain different genes

B different genes are switched on during development

C the genetic code is different in each cell

D they have different chromosomes.

26. A deficiency of Vitamin D in humans leads to rickets as a result of poor absorption of

A nitrate

B calcium

C iron

D phosphate.

27. Which line of the table identifies correctly the hormones which stimulate the inter-conversion of glucose and glycogen?

	glucose → glycogen	*glycogen → glucose*
A	insulin	glucagon and adrenaline
B	glucagon and insulin	adrenaline
C	adrenaline and glucagon	insulin
D	adrenaline	glucagon and insulin

28. Which line in the table describes body temperature in endotherms and ectotherms?

	Regulated by metabolism	*Regulated by behaviour*	*Varies with the environmental temperature*
A	ectotherm	endotherm	ectotherm
B	endotherm	ectotherm	endotherm
C	endotherm	ectotherm	ectotherm
D	ectotherm	endotherm	endotherm

29. Chlorophyll contains the metal ion

A iron

B copper

C magnesium

D calcium.

30. A species of plant was exposed to various periods of light and dark, after which the flowering response was observed.

The results are shown below.

Light period (hours)	*Dark period* (hours)	*Response of plant*
4	20	Maximum flowering
4	10	Flowering
6	18	Maximum flowering
14	10	Flowering
18	9	No flowering
18	6	No flowering
18	10	Flowering

What appears to be the critical factor which stimulates flowering?

A A minimum dark period of 10 hours

B A light and dark cycle of at least 14 hours

C A maximum dark period of 10 hours

D A dark period of at least 20 hours

Candidates are reminded that the answer sheet MUST be returned INSIDE the front cover of this answer book.

[Turn over

DO NOT WRITE IN THIS MARGIN

Marks

SECTION B

All questions in this section should be attempted.

1. Two magnified unicellular organisms are shown in the diagrams.

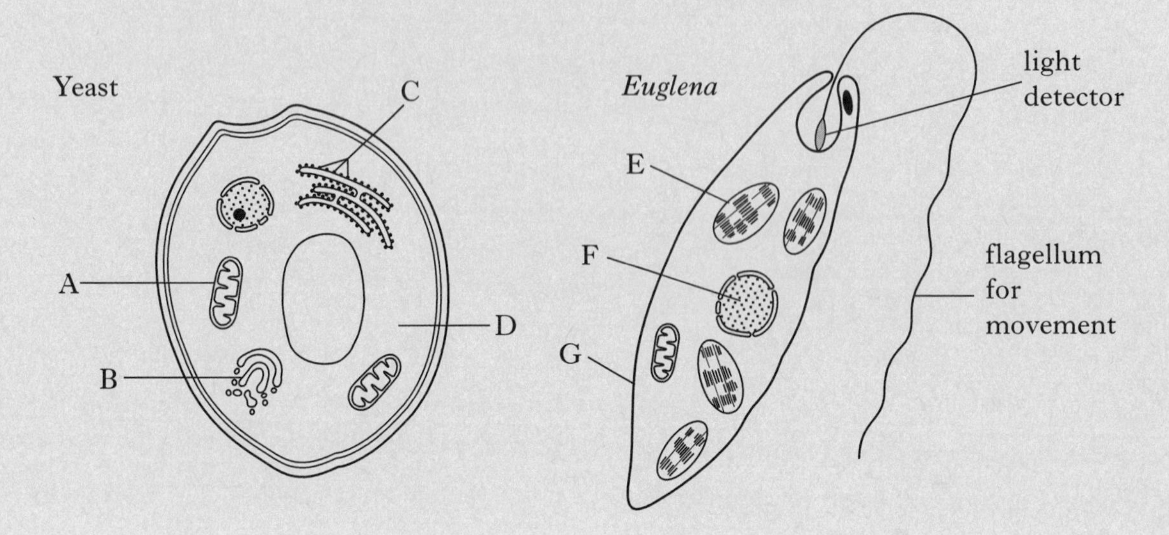

(*a*) (i) Name the **two** chemical components of structure G.

1 ______________________________

2 ______________________________ **1**

(ii) Complete the table by inserting letters from the diagrams to show where each process takes place.

Process	*Letter*
Glycolysis	
Transcription	

2

DO NOT WRITE IN THIS MARGIN

1. (continued) *Marks*

(*b*) What evidence from the diagram supports the statement that yeast cells secrete enzymes?

____ **1**

(*c*) *Euglena* lives in pond water. Explain how the structure of *Euglena* shown in the diagram allows it to photosynthesise efficiently.

____

____

____ **2**

[Turn over

DO NOT WRITE IN THIS MARGIN

Marks

2. An investigation was carried out into the effects of osmosis on beetroot tissue.

Pieces of beetroot were immersed in salt solutions of different concentration for one hour.

The results are shown in the table.

Concentration of salt solution (M)	*Mass of beetroot at start* (g)	*Mass of beetroot after 1 hour* (g)	*Percentage change in mass* (%)
0·05	4·0	4·8	+20
0·10	3·5	4·2	+20
0·20	4·4	4·7	+7
0·25	3·7	3·7	0
0·35	3·9	3·4	−13
0·40	3·5	2·8	−20

(*a*) On the grid, plot a line graph to show the percentage change in mass of the beetroot pieces against concentration of salt solution.

(Additional graph paper, if required, may be found on page 36.)

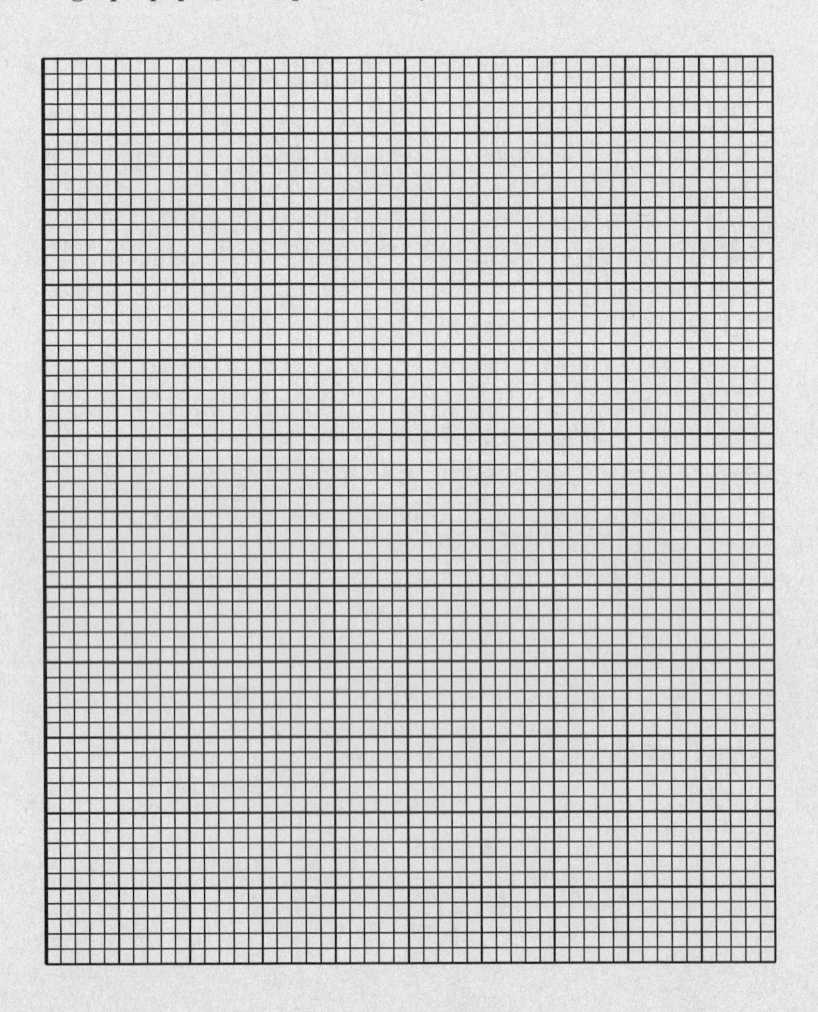

2

DO NOT WRITE IN THIS MARGIN

Marks

2. (continued)

(*b*) (i) Identify a concentration of salt solution used in the investigation that is hypertonic to the beetroot cell sap.

Explain your choice.

________________ M

Explanation ____

____

____ **1**

(ii) What term describes the condition of a plant cell after immersion for one hour in a 1·0 M salt solution?

____ **1**

(*c*) From the information given, why was it good experimental practice to use percentage change in mass when comparing results?

____

____ **1**

(*d*) Predict the percentage change in mass of a piece of beetroot immersed in 0·45 M salt solution for one hour.

________________ % **1**

(*e*) In setting up this investigation, variables were controlled to ensure that the results obtained would be valid.

Identify **one** variable related to the salt solutions and **one** variable related to the beetroot tissue which must be controlled.

Salt solutions ____ **1**

Beetroot tissue _______________________________________ **1**

[Turn over

DO NOT WRITE IN THIS MARGIN

Marks

3. (*a*) The diagram shows the role of the cytochrome system in aerobic respiration.

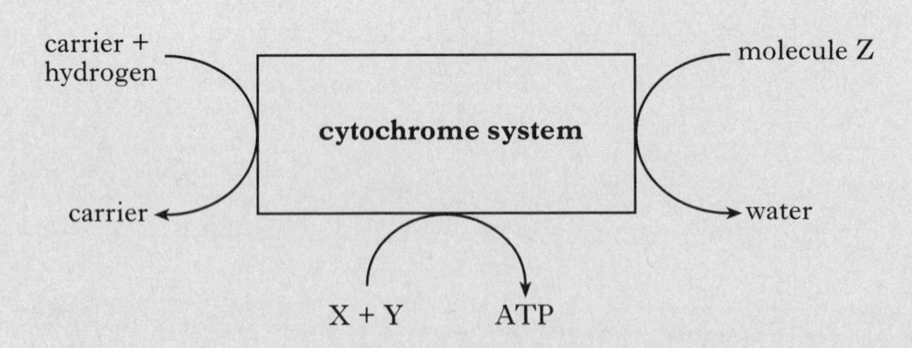

(i) State the exact location of the cytochrome system in a cell.

______________________________ **1**

(ii) Name the carrier that brings the hydrogen to the cytochrome system.

______________________________ **1**

(iii) Name molecules X, Y and Z.

X _______________ Y _______________ **1**

Z _______________ **1**

(*b*) The graph shows the effect of different conditions on the uptake of nitrate ions by barley roots.

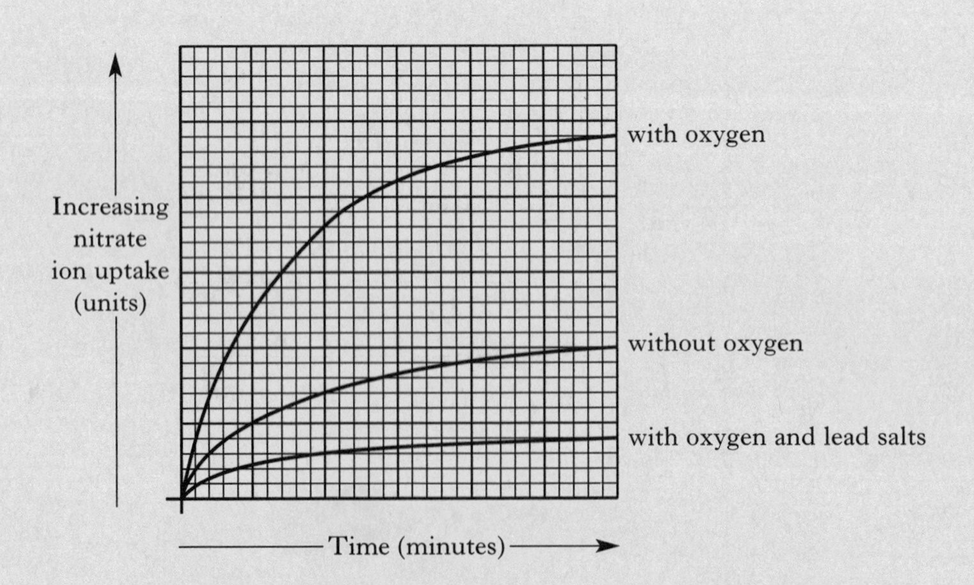

(i) State the importance of nitrate for the growth of barley plants.

____

______________________________ **1**

DO NOT WRITE IN THIS MARGIN

Marks

3. (*b*) (continued)

(ii) Explain why the uptake of nitrate ions is greater when oxygen is present.

____

____

____ **2**

(iii) Explain the effect of lead salts on nitrate ion uptake.

____

____

____ **2**

[Turn over

Marks

4. (*a*) The replication of part of a DNA molecule is represented in the diagram.

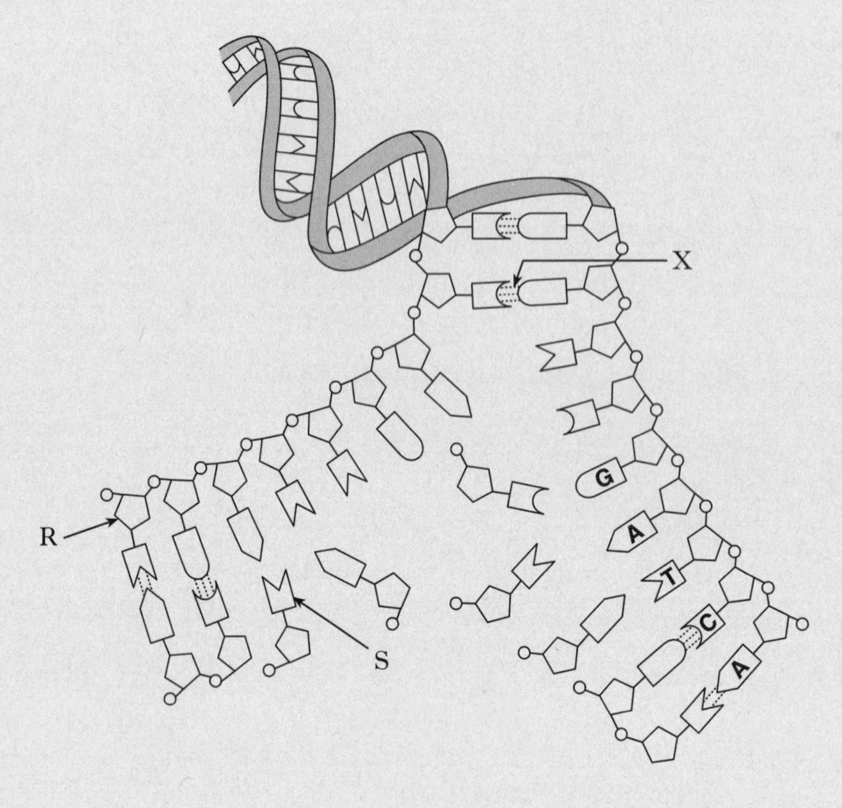

(i) Name the nucleotide component R and the base S.

R ______________________________ **1**

S ______________________________ **1**

(ii) Name the type of bond labelled X.

X ______________________________ **1**

(*b*) Explain why DNA replication must take place before a cell divides.

____

____

______________________________ **1**

DO NOT WRITE IN THIS MARGIN

Marks

4. (continued)

(*c*) Part of one strand of a DNA molecule used to make mRNA contains the following base sequence.

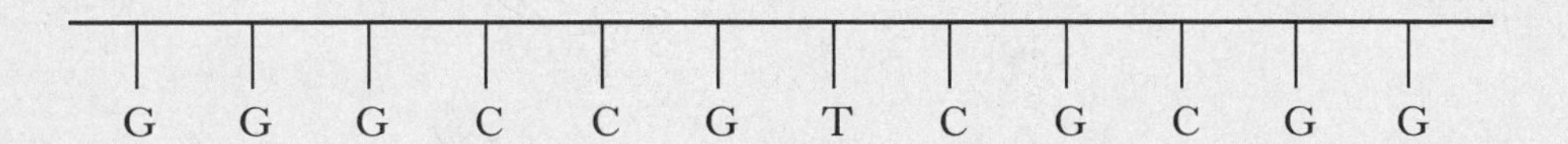

The table shows the names of six amino acids together with some of their mRNA codons.

Amino acid	*mRNA codon (s)*	
Glycine	GGG	GGC
Serine	UCG	AGC
Proline	CCG	CCC
Arginine	CGG	
Alanine	GCC	
Threonine	ACG	

(i) Use the information to give the order of amino acids coded for by the DNA base sequence.

______________________________ **1**

(ii) What name is given to a part of a DNA molecule which carries the code for making **one** protein?

______________________________ **1**

(*d*) Name the molecules that transport amino acids to the site of protein synthesis.

______________________________ **1**

(*e*) Complete the diagram below which shows information about protein classification.

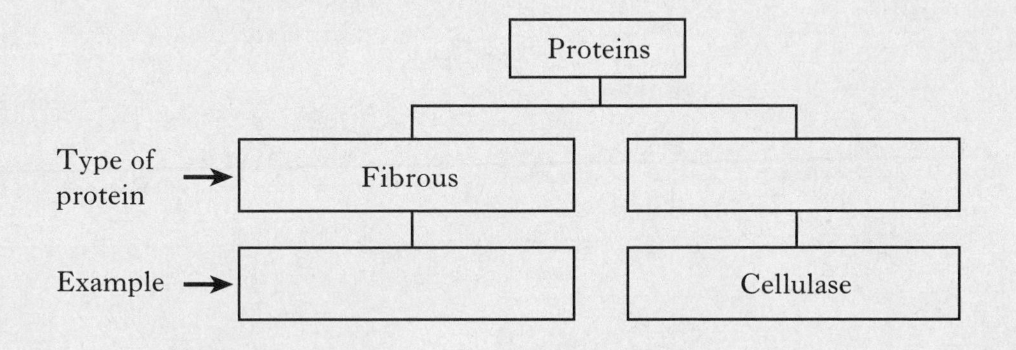

1

[Turn over

DO NOT WRITE IN THIS MARGIN

Marks

5. (*a*) The diagram shows two chromosomes and their appearance after a mutation has occurred.

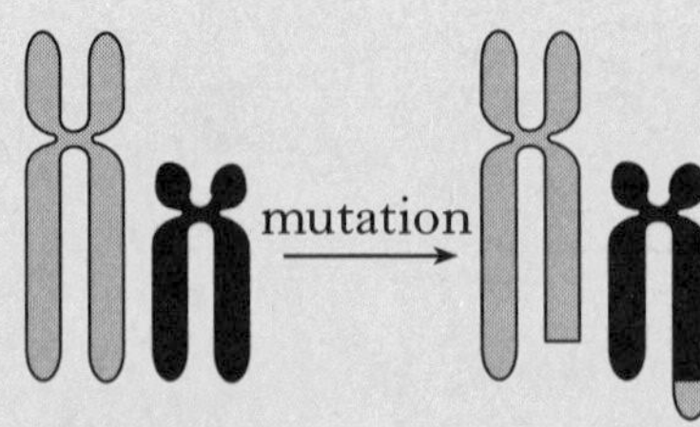

(i) Name this type of chromosome mutation.

______________________________ **1**

(ii) Name a mutagenic agent which could have caused this mutation.

______________________________ **1**

(*b*) Individuals with Down's Syndrome have 47 chromosomes in each cell instead of 46.

How does this change in chromosome number arise?

______________________________ **1**

(*c*) The diagram shows part of the normal amino acid sequence of an enzyme involved in a metabolic pathway. It also shows the altered sequence obtained after a gene mutation had occurred.

Normal amino acid sequence	His	Leu	Val	Glu	Ala	Leu	Tyr	Phe
Altered amino acid sequence	His	Leu	Met	Tyr	Met	Cys	Ileu	Ser

(i) Name a type of gene mutation which could have produced this altered amino acid sequence.

______________________________ **1**

(ii) Explain the effect this gene mutation would have on the metabolic pathway in which this enzyme is involved.

____

______________________________ **1**

(*d*) The DNA in one cell consists of 40 000 genes. During DNA replication, random mutations occur at the rate of one altered gene in every 625.

Calculate the average number of mutations which will occur during the full replication of this cell's DNA.

Space for working

______________________________ **1**

DO NOT WRITE IN THIS MARGIN

Marks

6. (*a*) A gene from a jellyfish can be inserted into a bacterial plasmid using a genetic engineering procedure.

Some of the stages involved are shown in the diagram.

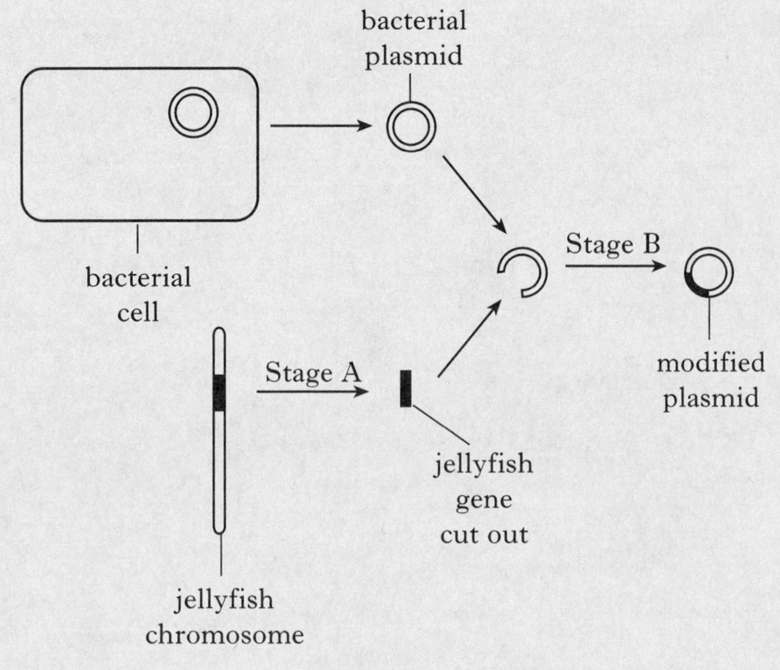

(i) Give **one** method which could be used for locating the gene in the jellyfish chromosome.

________________________________ **1**

(ii) Name the enzymes involved in the following stages of the genetic engineering procedure.

1 Cutting the jellyfish gene out of its chromosome (Stage A).

________________________________ **1**

2 Sealing the jellyfish gene into the bacterial plasmid (Stage B).

________________________________ **1**

(*b*) Name **one** human hormone that is manufactured by genetically engineered bacteria.

________________________________ **1**

[Turn over

DO NOT WRITE IN THIS MARGIN

Marks

7. (*a*) The flow diagram shows some stages in the regulation of water concentration of blood in mammals.

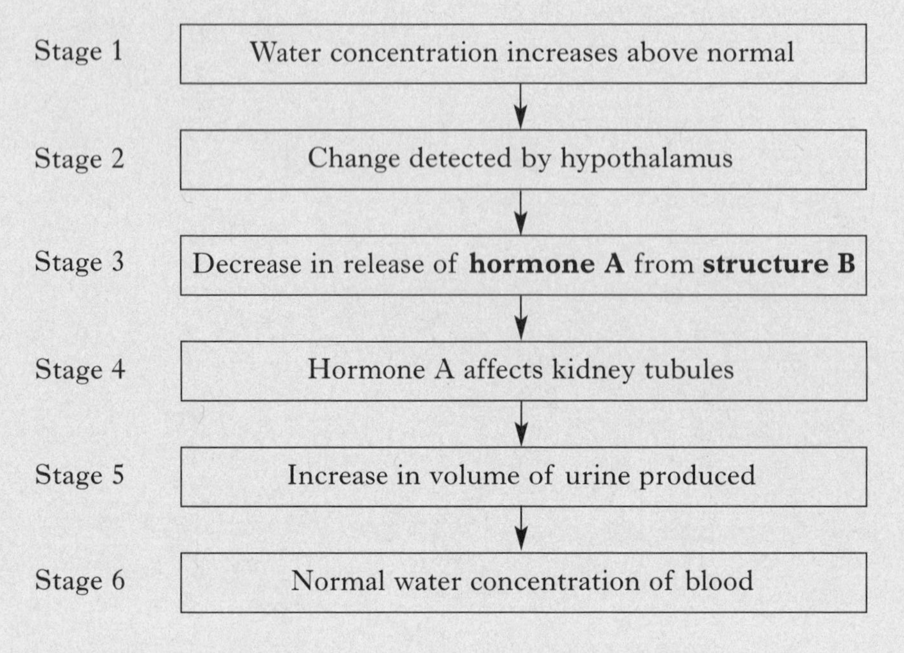

Stage 1
Water concentration increases above normal
Stage 2
Change detected by hypothalamus
Stage 3
Decrease in release of **hormone A** from **structure B**
Stage 4
Hormone A affects kidney tubules
Stage 5
Increase in volume of urine produced
Stage 6
Normal water concentration of blood

(i) Give **one** reason why the water concentration of the blood could increase above normal at stage 1.

______________________________ **1**

(ii) 1 Name hormone A ______________________________ **1**

2 Name structure B ______________________________ **1**

(iii) Describe how hormone A is transported to the kidney tubules.

______________________________ **1**

(iv) Describe the effect of hormone A on the kidney tubules.

____

______________________________ **1**

(v) What would be the effect of a decrease in hormone A on the concentration of salts in the urine?

______________________________ **1**

DO NOT WRITE IN THIS MARGIN

Marks

7. (continued)

(*b*) Salmon migrate between sea water and fresh water.

The table contains statements about osmoregulation in salmon.

For each statement, tick (✓) **one** box to show whether the statement is true for a salmon living in sea water or in fresh water.

Statement	*Sea water*	*Fresh water*
Salmon drinks a large volume of water		
Salmon produces a large volume of urine		
Chloride secretory cells pump out ions		
Salmon gains water by osmosis		

2

[Turn over

DO NOT WRITE IN THIS MARGIN

Marks

8. An experiment was carried out to investigate the growth of pea plants kept in a high light intensity following germination.

The graph shows the average dry mass and average shoot length of the pea plants.

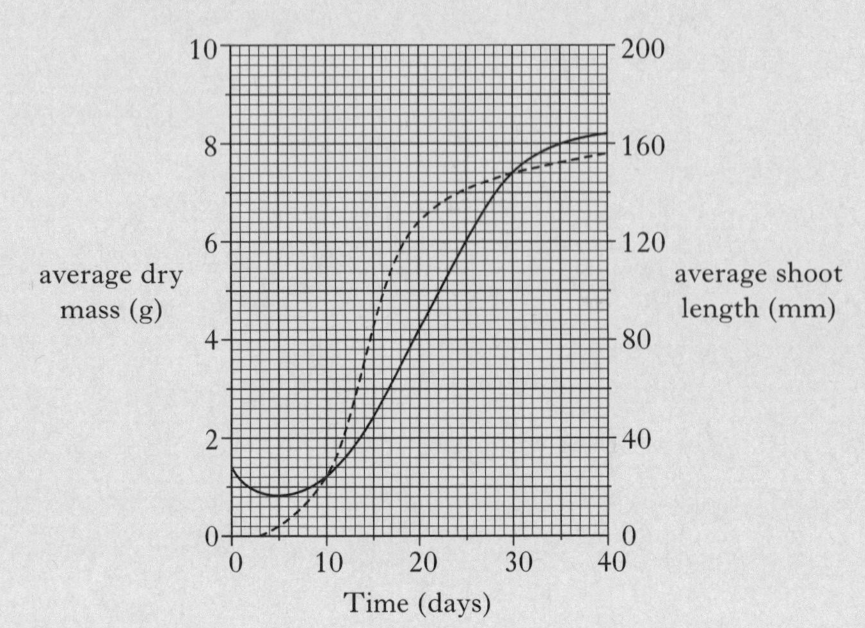

(*a*) (i) From the graph, how many days does it take for the shoot to emerge from the seed?

_______ days **1**

(ii) During which 5 day period is there the greatest increase in average shoot length? Tick (✓) one box.

Day 5–10	Day 10–15	Day 15–20	Day 20–25	Day 25–30

1

(iii) Explain the changes in average dry mass of the plants during the first fifteen days.

____

____

______________________________ **2**

(iv) Explain why measurement of average shoot length alone may not provide a reliable estimate of plant growth.

____

____

______________________________ **1**

DO NOT WRITE IN THIS MARGIN

8. (*a*) (continued)

Marks

(v) On day 30 the shoots made up 50% of the average dry mass of the plants. Calculate the average dry mass of the shoots per millimetre.

Space for calculation

____________ g per mm **1**

(*b*) The experiment was repeated with pea plants kept in the dark.

Complete the following to show how the results on day 15 would compare with the results obtained from plants grown in the light.

In each case, underline **one** alternative and give a reason to justify your choice.

(i) Average dry mass would be { greater. / less. / the same. }

Reason ________________________________ **1**

(ii) Average shoot length would be { greater. / less. / the same. }

Reason ________________________________ **1**

(*c*) The grid shows some of the effects of the plant growth substances Indole Acetic Acid (IAA) and Gibberellic Acid (GA) on the growth and development of plants.

A stimulates α-amylase production in barley grains	B promotes the formation of fruit	C inhibits leaf abscission
D causes apical dominance	E involved in phototropism	F breaks dormancy of buds

(i) Use **all** the letters from the grid to complete the table to show which effects are caused by IAA and which are caused by GA.

Effects caused by IAA	*Effects caused by GA*

3

(ii) Give **one** practical application of plant growth substances.

____

________________________________ **1**

Marks

9. The diagram represents a section through a woody twig with an area enlarged to show the xylem vessels present.

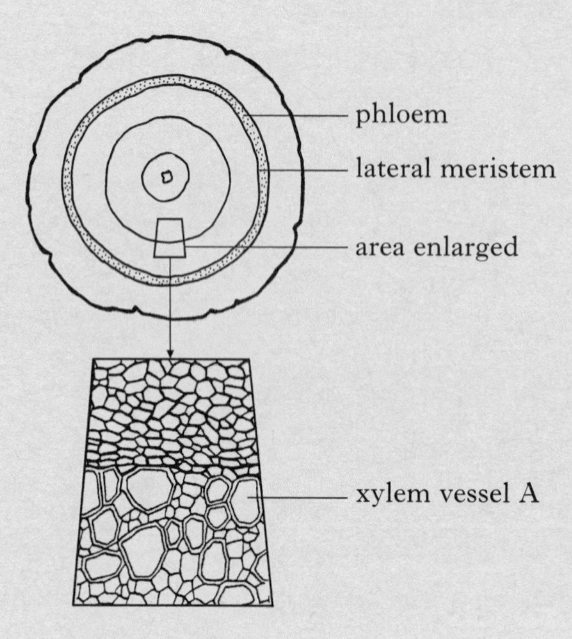

(*a*) Name the lateral meristem shown in the diagram.

____ **1**

(*b*) Explain how the appearance of xylem vessel A indicates that it was formed in the spring.

____

____

____ **1**

(*c*) What name is given to the area of the woody twig section that represents the xylem tissue growth occurring in one year?

____ **1**

DO NOT WRITE IN THIS MARGIN

Marks

10. Duckweed (*Lemna*) is a hydrophyte that has leaf-like structures which float on the surface of pondwater.

Some *Lemna* plants are shown in the diagram together with a magnified vertical section through one of the floating leaf-like structures.

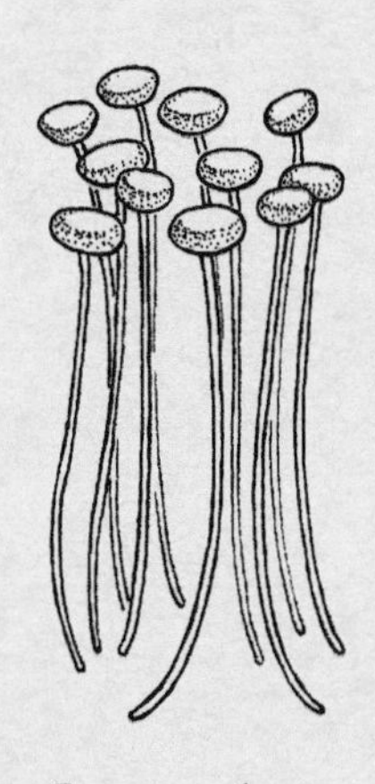

Lemna plants

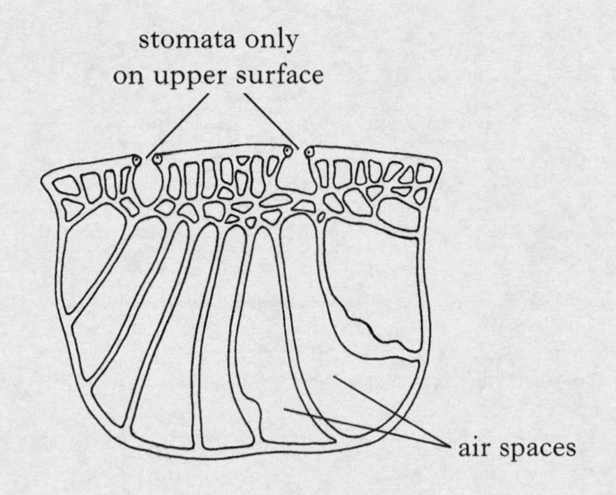

magnified vertical section

(*a*) Complete the table to describe the effect of each adaptation in *Lemna*.

Adaptation	*Effect*
Many large air spaces	
Stomata on upper surface	

1

1

(*b*) What term describes a plant that is adapted to live in a hot, dry habitat?

______________________________ 1

[Turn over

Marks

11. An investigation was carried out into the effects of competition when two species of flour beetle, *Tribolium confusum* and *Tribolium castaneum*, were kept together in a container with a limited food supply.

Tribolium beetles can be infected by a parasite which causes disease.

Graph 1 shows the numbers of the two species over the period of time in the absence of the parasite.

Graph 2 shows the effect of the presence of the parasite on the beetle numbers.

Graph 1 Parasite absent

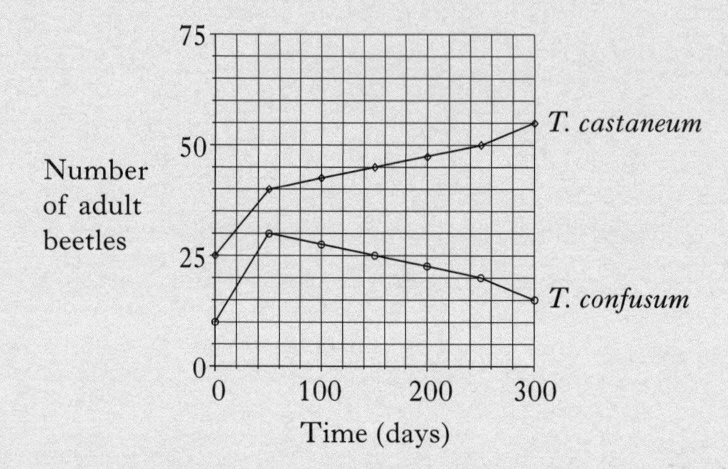

Graph 2 Parasite present

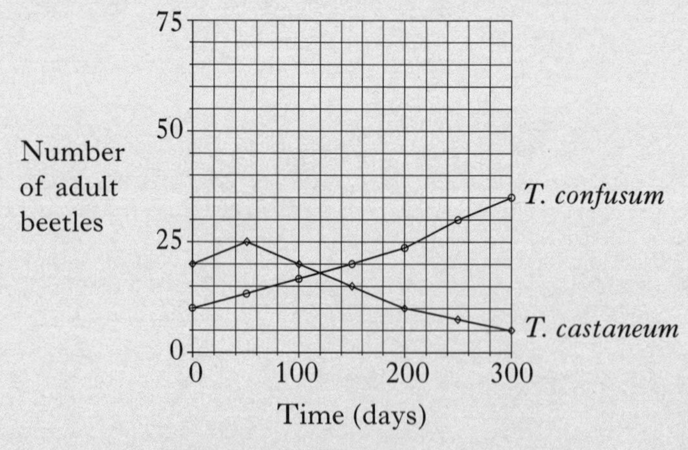

(*a*) Use values from **Graph 1** to describe how the numbers of *T. confusum* change over the period of the investigation.

____

____

________________________________ **2**

DO NOT WRITE IN THIS MARGIN

Marks

11. (continued)

(*b*) From **Graph 1**, express as the simplest whole number ratio the population size of *T. confusum* to *T. castaneum* at 250 days.

Space for calculation

T. confusum : *T. castaneum* ______ : ______ **1**

(*c*) From **Graph 2**, calculate the percentage increase in the *T. confusum* population over the 300 days of the investigation.

Space for calculation

______ % increase **1**

(*d*) Suggest an explanation for the improved growth of the *T. confusum* population in the presence of the parasite.

______

______

______ **2**

(*e*) From the information in the graphs, suggest an improvement to the design of the investigation.

______

______ **1**

(*f*) Certain factors may affect the numbers of beetles in this investigation.

Place ticks in the table to show whether each factor would have a density-dependent effect or a density-independent effect.

Factors	*Density-dependent*	*Density-independent*
Presence of disease causing parasites		
Availability of food		
Extreme temperature		

2

DO NOT WRITE IN THIS MARGIN

Marks

12. The diagram shows parts of the chromosome in the bacterium *E. coli*. The list has three molecules involved in the genetic control of lactose metabolism.

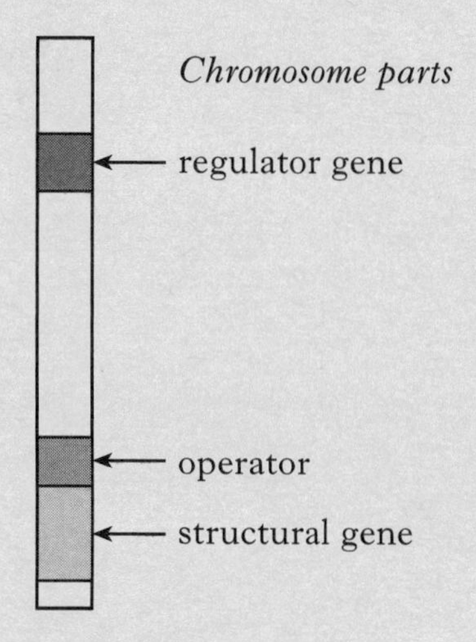

List of molecules

lactose-digesting enzyme
repressor
inducer

(*a*) Complete the table by writing **True** or **False** in each of the spaces provided.

Statement	*True/False*
The repressor can bind to the operator.	
The structural gene codes for the repressor.	
The inducer can bind to the repressor.	
The regulator gene codes for the lactose-digesting enzyme.	

2

(*b*) Name the inducer molecule.

______________________________________ **1**

(*c*) Give **one** advantage to *E. coli* of having this type of genetic control system.

____

______________________________________ **1**

DO NOT WRITE IN THIS MARGIN

SECTION C

Marks

Both questions in this section should be attempted.

Note that each question contains a choice.

Questions 1 and 2 should be attempted on the blank pages which follow.

Supplementary sheets, if required, may be obtained from the invigilator.

Labelled diagrams may be used where appropriate.

1. Answer **either** A **or** B.

A. Give an account of meiosis under the following headings:

(i) first meiotic division; **6**

(ii) second meiotic division; **2**

(iii) importance of meiosis. **2**

(10)

OR

B. Give an account of the evolution of new species under the following headings:

(i) isolating mechanisms; **4**

(ii) effects of mutations and natural selection. **6**

(10)

In question 2, ONE mark is available for coherence and ONE mark is available for relevance.

2. Answer **either** A **or** B.

A. Give an account of chloroplast structure in relation to the location of the stages of photosynthesis and describe the separation of photosynthetic pigments by chromatography. **(10)**

OR

B. Give an account of the nature of viruses and the production of more viruses. **(10)**

[*END OF QUESTION PAPER*]

SPACE FOR ANSWERS

[BLANK PAGE]

FOR OFFICIAL USE

Total for Sections B and C

X007/301

NATIONAL QUALIFICATIONS 2005

WEDNESDAY, 18 MAY 1.00 PM – 3.30 PM

BIOLOGY HIGHER

Fill in these boxes and read what is printed below.

Full name of centre

Town

Forename(s)

Surname

Date of birth
Day Month Year

Scottish candidate number

Number of seat

SECTION A—Questions 1–30 (30 marks)

Instructions for completion of Section A are given on page two.

SECTIONS B AND C (100 marks)

1. (a) All questions should be attempted.
   (b) It should be noted that in **Section C** questions 1 and 2 each contain a choice.
2. The questions may be answered in any order but all answers are to be written in the spaces provided in this answer book, and must be written clearly and legibly in ink.
3. Additional space for answers and rough work will be found at the end of the book. If further space is required, supplementary sheets may be obtained from the invigilator and should be inserted inside the **front** cover of this book.
4. The numbers of questions must be clearly inserted with any answers written in the additional space.
5. Rough work, if any should be necessary, should be written in this book and then scored through when the fair copy has been written. If further space is required a supplementary sheet for rough work may be obtained from the invigilator.
6. Before leaving the examination room you must give this book to the invigilator. If you do not, you may lose all the marks for this paper.

LIB X007/301 6/3/13320

Read carefully

1 Check that the answer sheet provided is for **Biology Higher (Section A)**.

2 Check that the answer sheet you have been given has **your name**, **date of birth**, **SCN** (Scottish Candidate Number) and **Centre Name** printed on it.

Do not change any of these details.

3 If any of this information is wrong, tell the Invigilator immediately.

4 If this information is correct, **print** your name and seat number in the boxes provided.

5 Use **black** or **blue ink** for your answers. **Do not use red ink**.

6 The answer to each question is **either** A, B, C or D. Decide what your answer is, then put a horizontal line in the space provided (see sample question below).

7 There is **only one correct** answer to each question.

8 Any rough working should be done on the question paper or the rough working sheet, **not** on your answer sheet.

9 At the end of the exam, put the **answer sheet for Section A inside the front cover of this answer book**.

Sample Question

The apparatus used to determine the energy stored in a foodstuff is a

A respirometer

B calorimeter

C klinostat

D gas burette

The correct answer is **B**—calorimeter. The answer **B** has been clearly marked with a horizontal line (see below).

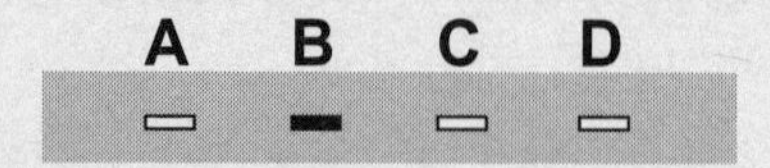

Changing an answer

If you decide to change your answer, cancel your first answer by putting a cross through it (see below) and fill in the answer you want. The answer below has been changed to **B**.

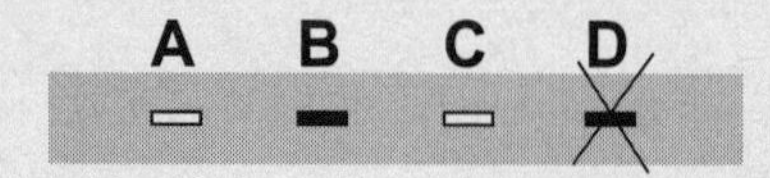

If you then decide to change back to an answer you have already scored out, put a tick (✓) to the **right** of the answer you want, as shown below:

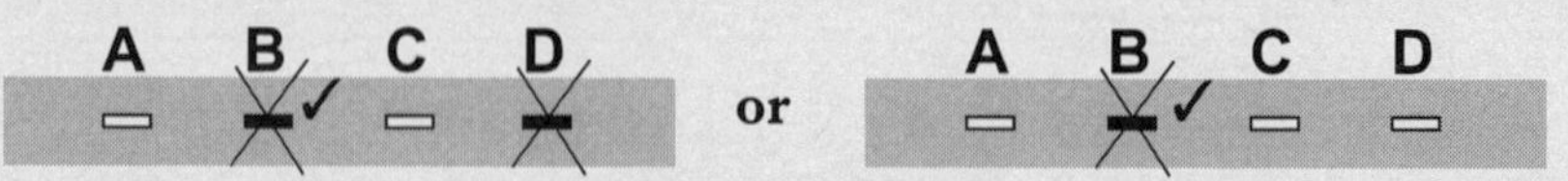

SECTION A

All questions in this section should be attempted.

Answers should be given on the separate answer sheet provided.

1. When a red blood cell is immersed in a hypertonic solution it will

A shrink

B become flaccid

C burst

D become turgid.

2. The diagram below represents some of the structures present in a plant cell.

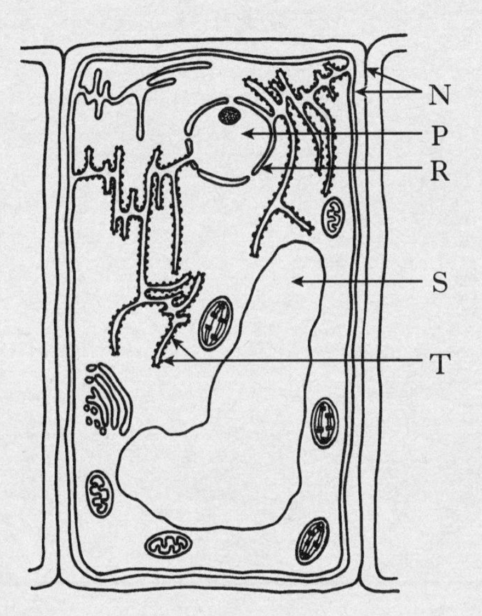

Which line in the table matches the structures with the materials of which they are mainly composed?

	Materials	
	protein and phospholipid	*nucleic acid and protein*
A	R	P
B	R	S
C	T	R
D	R	N

3. Which of the following is a structural carbohydrate?

A Glucose

B Starch

C Glycogen

D Cellulose

4. The graph illustrates the effects of light intensity, temperature and carbon dioxide (CO_2) concentration on the rate of photosynthesis.

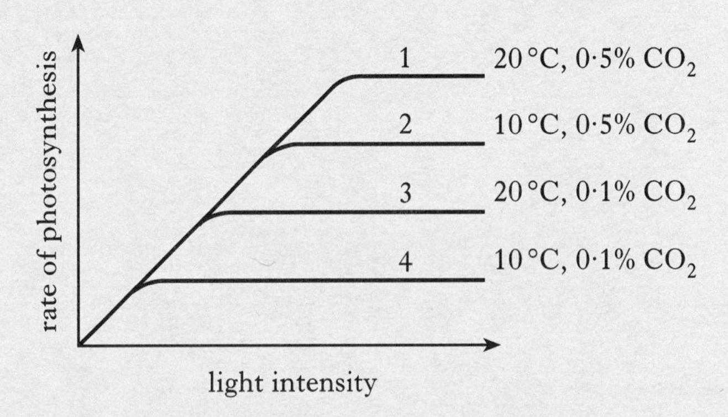

Which of the following pairs of lines in the graph suggest that carbon dioxide is acting as a limiting factor?

A 2 and 4

B 2 and 3

C 1 and 4

D 1 and 2

5. Which of the following elements is essential to the formation of chlorophyll?

A Potassium

B Magnesium

C Copper

D Calcium

6. Which of the following is composed of protein?

A Nucleotide

B Glycogen

C Antibody

D Polysaccharide

[Turn over

7. How many adenine molecules are present in a DNA molecule of 2000 bases, if 20% of the base molecules are cytosine?

A 200

B 300

C 400

D 600

8. Which of the following statements is true of all viruses?

A They have a protein-lipid coat and contain DNA.

B They have a protein-lipid coat and contain RNA.

C They have a protein coat and a nucleus.

D They have a protein coat and contain nucleic acid.

9. The genes of viruses are composed of

A either DNA or RNA

B DNA only

C RNA only

D enzymes and nucleic acids.

10. In infertility clinics, samples of semen are collected for testing.

The table below refers to the analysis of semen samples taken from five men.

Semen sample	*1*	*2*	*3*	*4*	*5*
Number of sperm in sample (millions/cm^3)	40	19	25	45	90
Active sperm (percent)	65	60	75	10	70
Abnormal sperm (percent)	30	20	90	30	10

A man is fertile if his semen contains at least 20 million sperm cells/cm^3 and at least 60% of the sperm cells are active and at least 60% of the sperm cells are normal.

The semen samples that were taken from infertile men are

A samples 3 and 4 only

B samples 2 and 4 only

C samples 2, 3 and 4 only

D samples 1, 2, 4 and 5 only.

11. Alleles can be described as

A opposite types of gamete

B different versions of a gene

C identical chromatids

D non-homologous chromosomes.

12. Which of the following defines linkage?

A Genes which are transferred from one chromosome pair to another

B Genes which are present on the same chromosome

C Genes which are transferred from one chromosome to its partner

D Genes which are present on different chromosomes

13. The table below shows the recombination frequency between genes on a chromosome.

Crossing over between genes	*Recombination frequency*
F and G	4%
F and J	6%
G and H	6%
H and J	4%

Use the information in the table to work out the order of genes on the chromosome.

The order of the genes is

A H G F J

B F G H J

C F G J H

D G H F J.

14. In *Drosophila*, white eye colour is a sex-linked recessive character. If a homozygous white-eyed female is crossed with a red-eyed male, what will be the phenotype of the first generation?

A All females will be white-eyed and all males red-eyed.

B All females will be red-eyed and all males white-eyed.

C All females will be red-eyed and 1 in 2 males will be white-eyed.

D 1 in 4 will be white-eyed irrespective of sex.

15. Which of the following may result in the presence of an extra chromosome in the cells of a human being?

A Non-disjunction

B Crossing over

C Segregation

D Inversion

16. Which of the following is an example of the result of natural selection?

A Modern varieties of potato have been produced from wild varieties.

B Ayrshire cows have been selected through breeding for milk production.

C Bacterial species have developed resistance to antibiotics.

D Varieties of tomato plants have resistance to fungal diseases through somatic fusion.

17. The dark variety of the peppered moth became common in industrial areas of Britain following the increase in the production of soot during the Industrial Revolution.

The increase in the dark form was due to

A dark moths migrating to areas which gave the best camouflage

B a change in the prey species taken by birds

C an increase in the mutation rate

D a change in selection pressure.

18. Which of the following is true of the kidneys of a salt-water bony fish?

A They have few large glomeruli.

B They have few small glomeruli.

C They have many large glomeruli.

D They have many small glomeruli.

19. The Soft Brome Grass and Long Beaked Storksbill are species of plant which grow on the grasslands of California. The Storksbill is a low-growing plant with a more extensive root system than the Soft Brome, but does not grow as tall as the Soft Brome.

Under which of the following conditions would the Storksbill become the more abundant species?

A Drought

B High soil moisture levels

C High light intensity

D Shade

20. Which of the following best describes habituation?

A The same escape response is performed repeatedly.

B The same response is always given to the same stimulus.

C A harmless stimulus ceases to produce a response.

D Behaviour is reinforced by regular repetition.

[Turn over

21. Hawks are predators which attack flocks of pigeons. The graph below shows how attack success by a hawk varies with the number of pigeons in a flock.

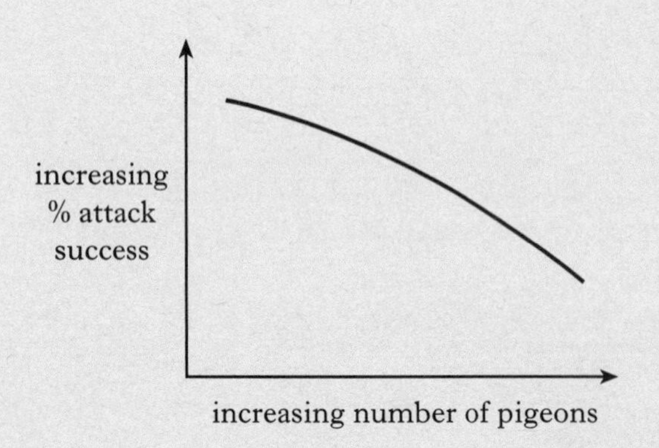

Which of the following statements could explain the observations shown in the graph?

A A hawk only needs to eat a small percentage of a large flock of prey.

B Co-operative hunting is more effective with small numbers of prey.

C A predator can be more selective when prey numbers increase.

D A hawk has difficulty focussing on one pigeon in a large flock.

22. Root tips are widely used for the study of mitosis because

A the cells are larger than other cells

B they contain many meristematic cells

C their nuclei have large chromosomes

D their nuclei stain easily.

23. The graphs below show the average yearly increase in height of girls and boys.

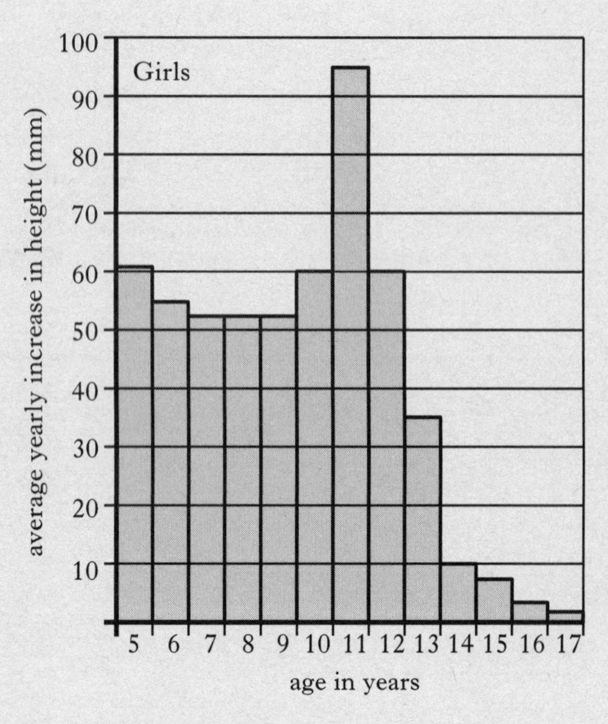

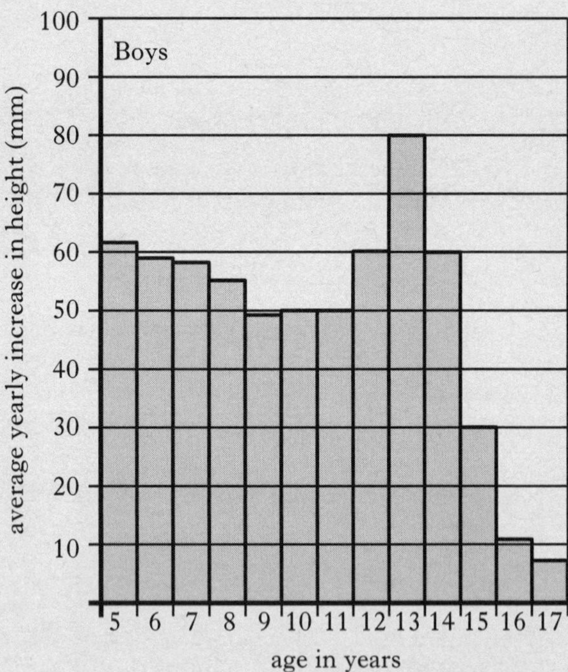

Which of the following statements is correct?

A The greatest average yearly increase for boys occurs one year later than the greatest average yearly increase for girls.

B Boys are still growing at seventeen but girls have stopped growing by this age.

C Between the ages of five and eight boys grow more than girls.

D There is no age when boys and girls show the same average yearly increase in height.

24. The following diagram shows an enzyme-controlled metabolic pathway.

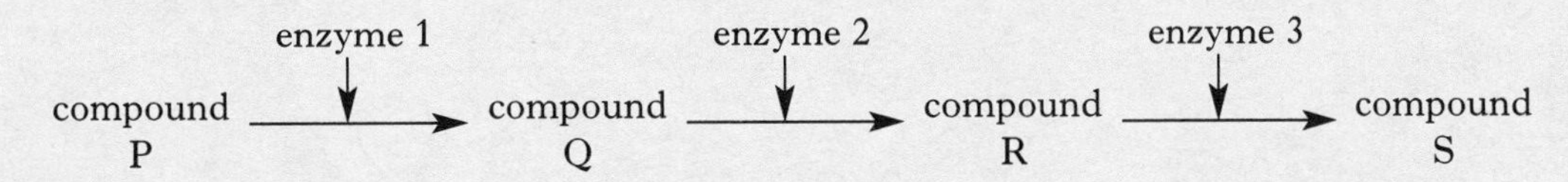

If enzyme 2 is inactivated (eg by adding an inhibitor) at time X shown in the graphs below, which graph predicts correctly the final concentration of compounds Q and R?

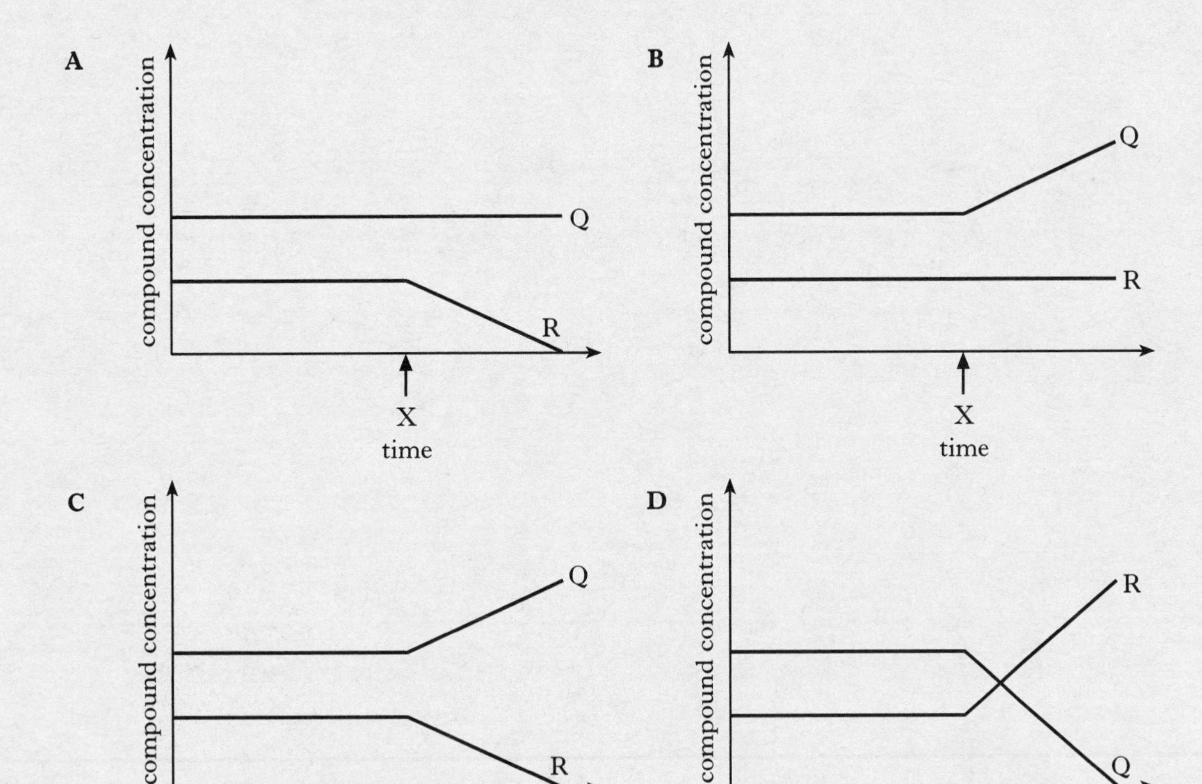

25. The table shows the results of an experiment carried out to study the effects of a plant growth substance on the roots of tomato plants.

Concentration of growth substance		*Control* 0 mg/litre	10^{-1} mg/litre
Average length of 20 roots	Before treatment	16 mm	16 mm
	After treatment	24 mm	20 mm

Which of the following states the effect of the plant growth substance on the lengths of the roots compared to the control treatment?

A 25 percent stimulation

B 50 percent stimulation

C 25 percent inhibition

D 50 percent inhibition

26. A short day plant is one which

A will flower only if the night length is less than the critical value

B will flower only if daylight is less than 12 hours

C will flower only if the hours of daylight are less than a critical value

D flowers only if the hours of daylight are more than a critical value.

27. A plant becomes etiolated when it

A grows in poor soil

B grows in the dark

C is treated with gibberellin

D has the apical bud removed.

 [Turn over

28. If the concentration of glucose in the blood of a healthy man or woman rises above normal, the pancreas produces

A more insulin but less glucagon

B more insulin and more glucagon

C less insulin but more glucagon

D less insulin and less glucagon.

29. If body temperature drops below normal, which of the following would result?

A Vasodilation of skin capillaries

B Vasoconstriction of skin capillaries

C Decreased metabolic rate

D Increased sweating

30. The diagram below represents a sandy coastal area. The sand deposits support various communities of plants.

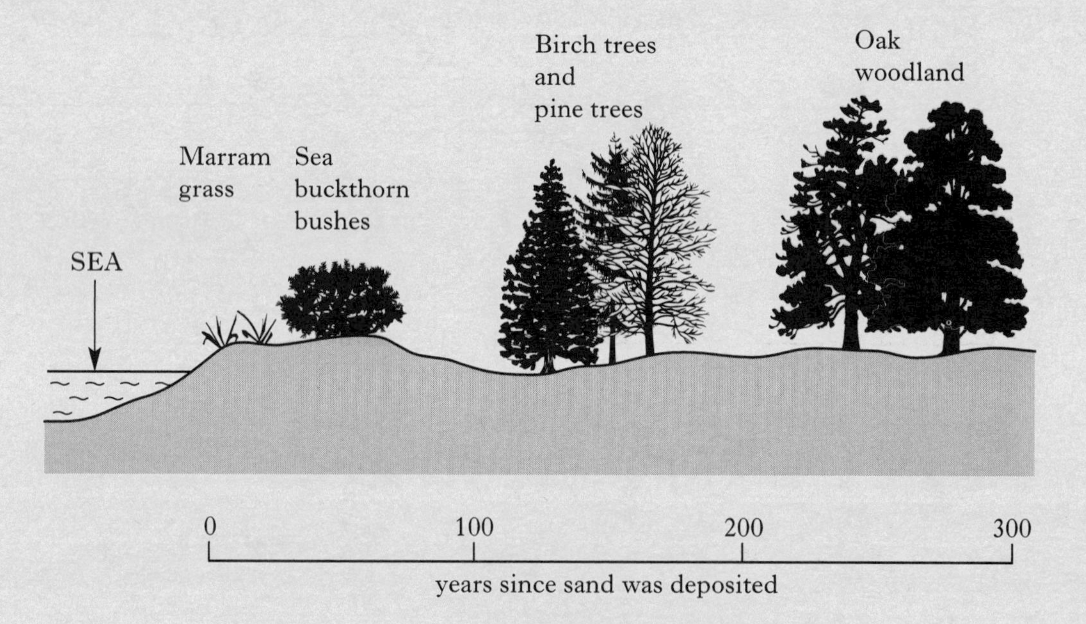

What term is used to describe the sequence of communities shown?

A Colonisation

B Climax

C Progression

D Succession

Candidates are reminded that the answer sheet MUST be returned INSIDE the front cover of this answer book.

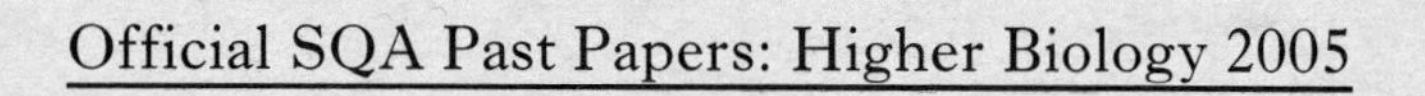

[Turn over for Section B on *Page ten*

DO NOT WRITE IN THIS MARGIN

Marks

SECTION B

All questions in this section should be attempted.

1. The diagram shows a mitochondrion from a human muscle cell.

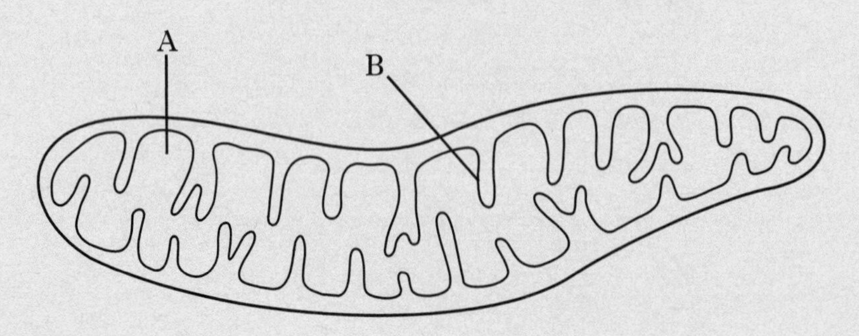

(*a*) Name regions A and B.

A ________________________________

B ________________________________ **1**

(*b*) The table shows some substances involved in respiration.

(i) Complete the table by inserting the number of carbon atoms present in each substance.

Substance	*Number of carbon atoms present*
Pyruvic acid	
Acetyl group	
Citric acid	

2

(ii) To which substance is the acetyl group attached before it enters the citric acid cycle?

________________________________ **1**

DO NOT WRITE IN THIS MARGIN

Marks

1. (continued)

(*c*) In region B, hydrogen is passed through a series of carriers in the cytochrome system as shown in the diagram below.

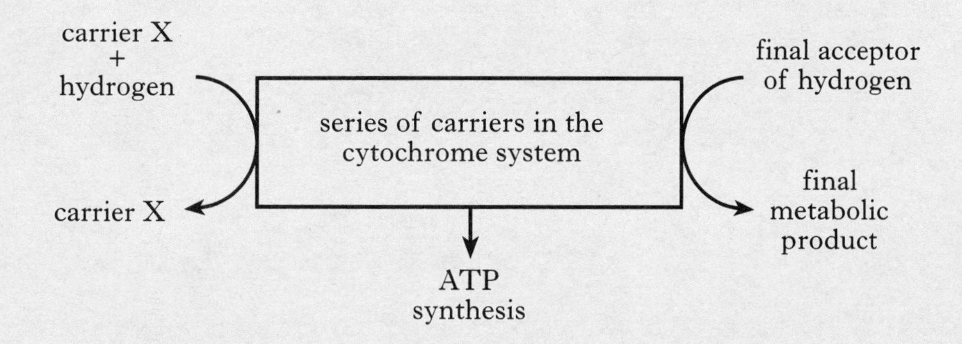

(i) Name carrier X.

______________________________ **1**

(ii) Name the final acceptor of hydrogen.

______________________________ **1**

(iii) Describe the importance of ATP in cells.

____

______________________________ **1**

(iv) The quantity of ATP present in the human body remains relatively constant yet ATP is continually being broken down.

Suggest an explanation for this observation.

____

____

______________________________ **1**

(*d*) Name the final metabolic product of **anaerobic** respiration in a muscle cell.

______________________________ **1**

[Turn over

Marks

2. The diagram shows two different types of blood cell involved in the defence of the human body.

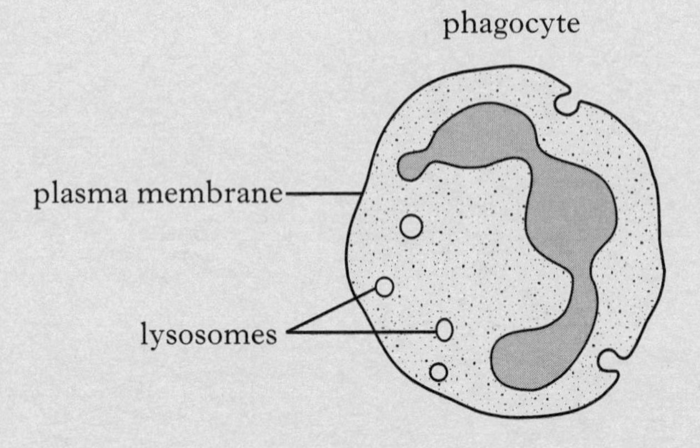

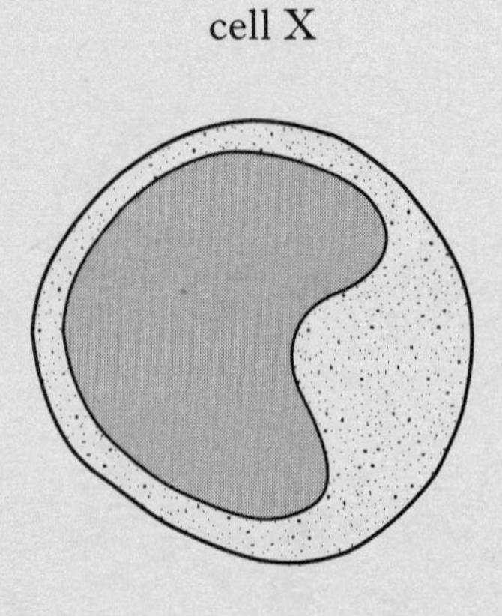

(*a*) Describe how the plasma membrane and the lysosomes of phagocytes are involved in helping to destroy bacteria.

(i) plasma membrane ______________________________________

______________________________________ **1**

(ii) lysosomes ______________________________________

______________________________________ **1**

(*b*) (i) Name cell X.

______________________ **1**

(ii) Explain how cell X may be involved in tissue rejection following a transplant operation.

____

____

______________________________________ **1**

(iii) What treatment is given to prevent tissue rejection?

______________________________________ **1**

DO NOT WRITE IN THIS MARGIN

Marks

3. In Dachshund dogs, the genes for hair texture and hair length are located on different chromosomes.

The allele for wire hair (**A**) is dominant to the allele for smooth hair (**a**).
The allele for short hair (**B**) is dominant to the allele for long hair (**b**).

Wire hair is **always** short so dogs with allele **A** are **always** short haired.

Two Dachshunds with the genotype **AaBb** were crossed.

(*a*) State the phenotype of the parents in this cross.

______________________________ **1**

(*b*) The grid shows all the genotypes of the offspring that may arise from this cross.

Complete the grid by adding the genotypes of the male and female gametes.

Male gametes

Female gametes				
	1 **AABB**	**AABb**	**AaBB**	**AaBb**
	AABb	2 **AAbb**	**AaBb**	**Aabb**
	AaBB	**AaBb**	3 **aaBB**	**aaBb**
	AaBb	**Aabb**	**aaBb**	4 **aabb**

1

(*c*) Complete the table below to give the phenotypes of the offspring indicated by the shaded boxes numbered 1 to 4 on the grid.

Box	*Phenotype*
1	
2	
3	
4	

2

(*d*) From the grid, calculate the expected ratio of the phenotypes of **all** the offspring from this cross.

Space for working

_____ wire short hair : _____ smooth short hair : _____ smooth long hair **1**

DO NOT WRITE IN THIS MARGIN

Marks

4. (*a*) The diagram shows the amino acid sequences of a fish hormone and two human hormones which may have evolved from it.

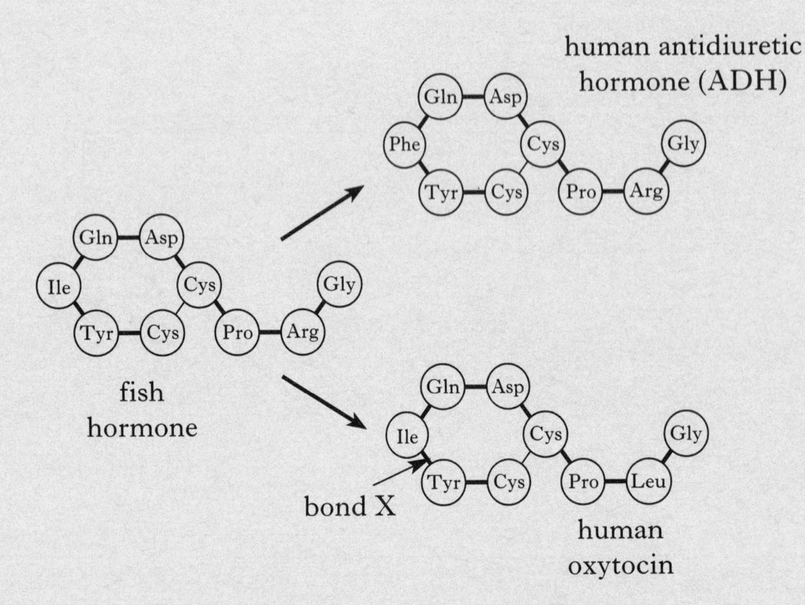

Amino acid key

Arg	arginine
Asp	aspartic acid
Cys	cysteine
Gln	glutamine
Gly	glycine
Ile	isoleucine
Leu	leucine
Phe	phenylalanine
Pro	proline
Tyr	tyrosine

(i) Name the type of bond represented by X.

______________________________ **1**

(ii) In the evolution of human oxytocin from the fish hormone, a gene mutation resulted in the amino acid arginine being replaced by leucine.

The table shows four of the mRNA codons for the amino acids arginine and leucine.

Codons for arginine	*Codons for leucine*
CGU	CUU
CGC	CUC
CGA	CUA
CGG	CUG

Name the type of gene mutation that occurred and justify your answer.

Type of gene mutation ______________________________ **1**

Justification ______________________________

______________________________ **1**

(iii) Describe the change in protein structure that occurred in the evolution of human antidiuretic hormone (ADH) from the fish hormone.

____

______________________________ **1**

DO NOT WRITE IN THIS MARGIN

Marks

4. (continued)

(*b*) Antidiuretic hormone (ADH) is involved in osmoregulation in humans.

(i) Name the gland that releases ADH.

______________________________ **1**

(ii) The graphs show the effects of increasing blood solute concentration and increasing blood volume on the plasma ADH concentration.

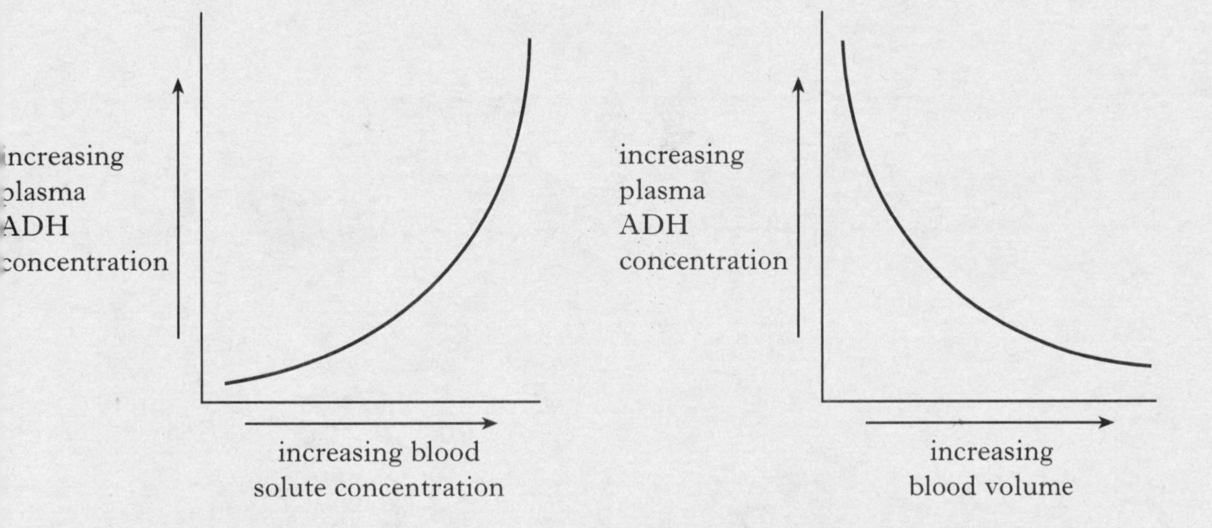

Use the information in the graphs to complete the table by using the terms "increases", "decreases" or "stays the same" to show the effect of various activities on the plasma ADH concentration.

Each term may be used **once**, **more than once** or **not at all**.

Activity	*Effect on plasma ADH concentration*
Drinking fresh water	
Sweating	
Eating salty food	
Severe bleeding	

2

(iii) Describe the effect that an increase in plasma ADH concentration has on the activity of kidney tubules.

____

______________________________ **1**

[Turn over

DO NOT WRITE IN THIS MARGIN

Marks

5. The diagram shows how an isolating mechanism can divide a population of one species into two sub-populations and then act as a barrier to prevent gene exchange between them.

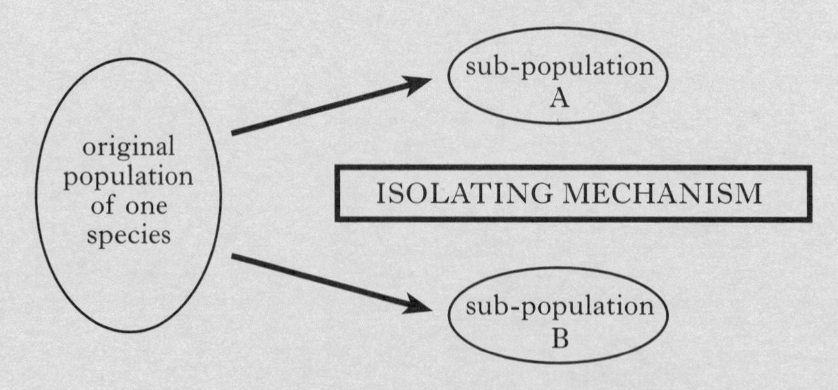

(*a*) Name **one** type of isolation that could prevent gene exchange between the two sub-populations.

____________________ **1**

(*b*) Over a long period of time, the gene pools of sub-populations A and B become different from each other.

(i) Explain how mutations and natural selection account for the differences.

1. Mutations ____________________

____

____________________ **1**

2. Natural selection ____________________

____

____

____________________ **2**

(ii) Eventually, sub-populations A and B may become two different species. What evidence would confirm that this had happened?

____

____________________ **1**

DO NOT WRITE IN THIS MARGIN

Marks

6. The list below contains terms related to genetic engineering and somatic fusion.

List of terms:

cellulase
gene probe
ligase
plasmid
protoplast
restriction endonuclease.

(*a*) Complete the table to match **each** of the following descriptions to the correct term from the above list.

Description	*Term*
Contains bacterial genes	
Cuts DNA into fragments	
Locates specific genes	
Removes plant cell walls	

2

(*b*) State the problem in plant breeding that is overcome by using the technique of somatic fusion.

____ **1**

[Turn over

DO NOT WRITE IN THIS MARGIN

Marks

7. (*a*) The diagram represents a plant with two regions magnified to show tissues involved in transport.

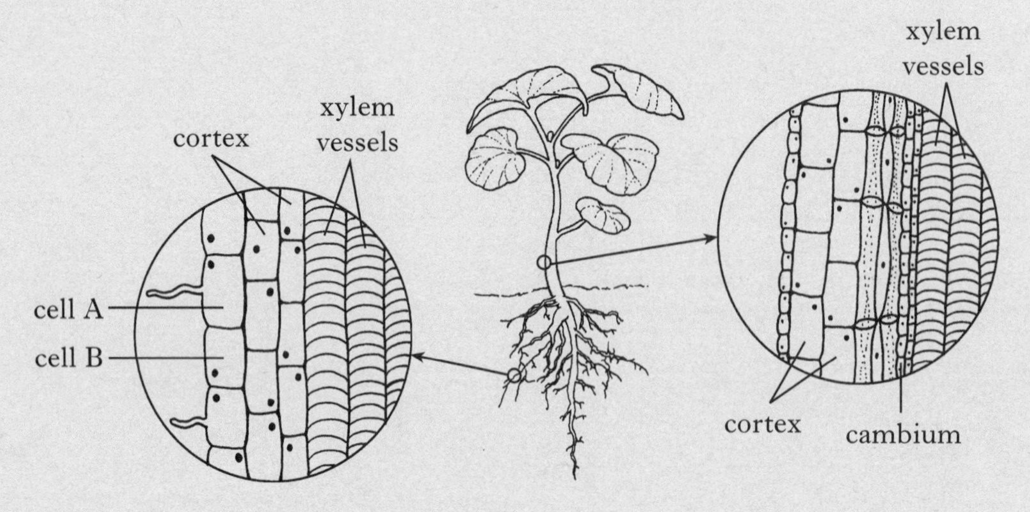

(i) Describe the process by which water moves into cell A. **1**

(ii) Cells A and B have a similar function.
Explain how the structure of cell A makes it better adapted to its function than cell B. **1**

(iii) Name the force that holds water molecules together as they travel up the xylem vessels. **1**

(iv) Cell division in the cambium produces new cells which then elongate and develop vacuoles.

Describe **two** further changes that take place in these cells as they differentiate into xylem vessels.

1 **1**

2 **1**

DO NOT WRITE IN THIS MARGIN

Marks

7. (continued)

(*b*) The diagrams show stomata in the lower epidermis of a leaf.

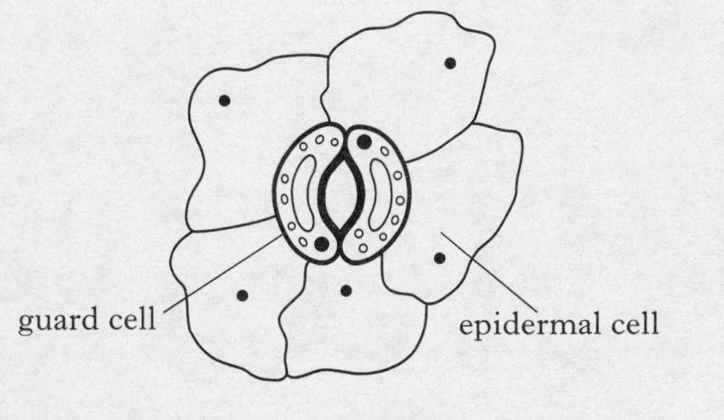

open stoma

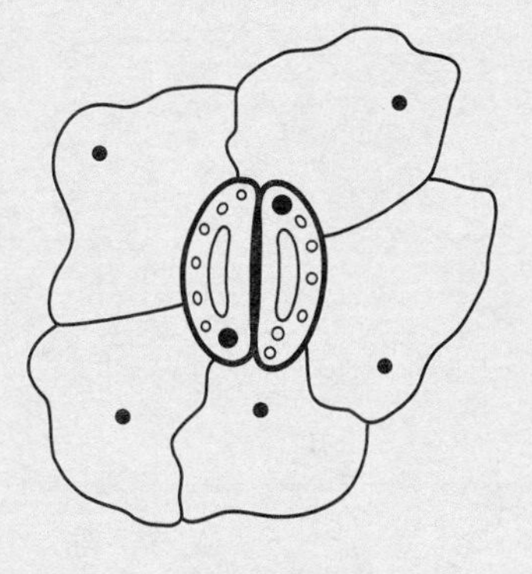

closed stoma

(i) In the following sentence, **underline** one of the alternatives in each pair to make the sentence correct.

Stomata close when water moves {into / out of} the guard cells and they become {more / less} turgid. **1**

(ii) What is the advantage to plants in having their stomata closed at night?

____ **1**

(*c*) The grid shows factors affecting the rate of transpiration from leaves.

A increased temperature	**B** increased wind speed	**C** increased humidity
D decreased temperature	**E** decreased wind speed	**F** decreased humidity

(i) Which **three** letters indicate the changes that would result in a decrease in the rate of transpiration?

Letters __________ , __________ and __________. **1**

(ii) The transpiration stream supplies plant cells with water for photosynthesis.

Give **one** other benefit to plants of the transpiration stream.

____ **1**

8. **Figure 1** shows how glycerate phosphate (GP) and ribulose bisphosphate (RuBP) are involved in the Calvin cycle.

Figure 1

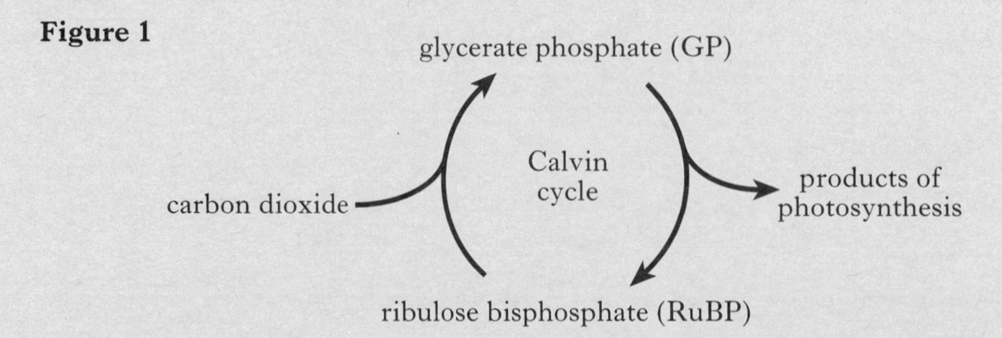

An investigation of the Calvin cycle was carried out in *Chlorella*, a unicellular alga.

Graph 1 shows the concentrations of GP and RuBP in *Chlorella* cells kept in an illuminated flask at 15 °C. The concentration of carbon dioxide in the flask was 0·05% for the first three minutes, then it was reduced to 0·005%.

Graph 1

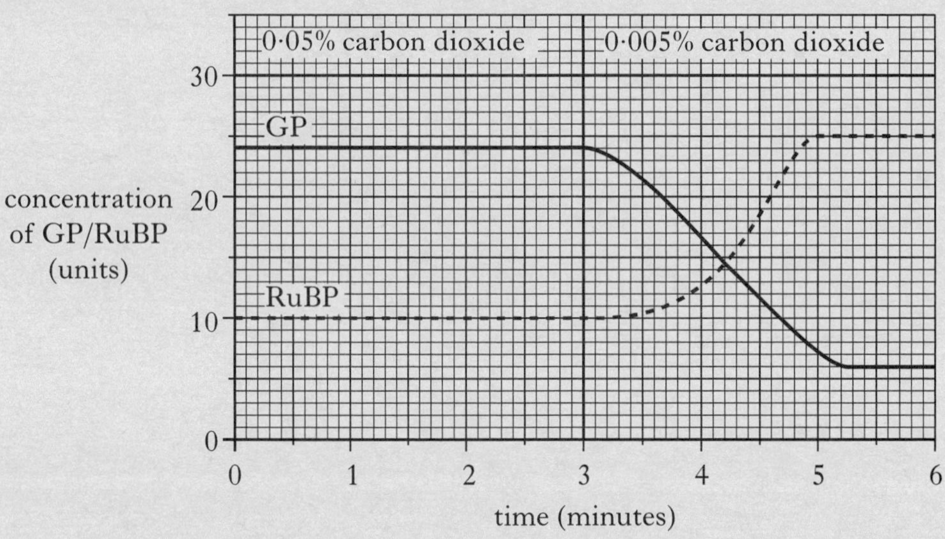

Graph 2 shows the rate of carbon dioxide fixation by *Chlorella* cells at various carbon dioxide concentrations.

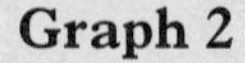
Graph 2

rate of carbon dioxide fixation ($mmol\,h^{-1}$)

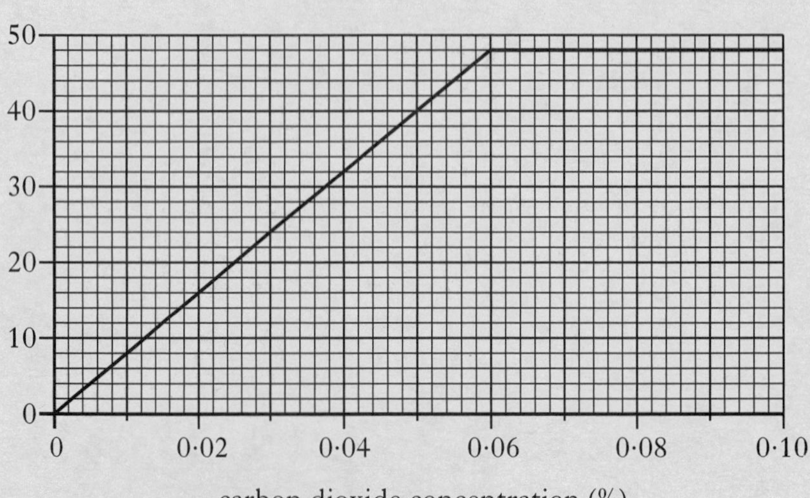

DO NOT WRITE IN THIS MARGIN

8. (continued) *Marks*

(*a*) (i) Use values from **Graph 1** to describe the changes in the RuBP concentration over the first six minutes.

____

____

______________________________ **2**

(ii) Use the information in **Figure 1** to explain the increase in RuBP concentration shown in **Graph 1** when the carbon dioxide concentration is decreased.

____

____

______________________________ **2**

(*b*) From **Graph 1**, calculate the percentage decrease in the concentration of GP from three to six minutes.

Space for calculation

__________ % **1**

(*c*) Use the terms "increase", "decrease" or "stay the same" to complete the sentence below. Each term may be used **once**, **more than once** or **not at all**.

If the carbon dioxide concentration was returned to 0·05% at 6 minutes,

the concentration of RuBP would ______________________________

and the concentration of GP would ______________________________. **1**

(*d*) From **Graph 2**, state the rate of carbon dioxide fixation by *Chlorella* at a carbon dioxide concentration of 0·01%.

__________ mmol h^{-1} **1**

(*e*) How many times greater is the rate of carbon dioxide fixation from 0 to 3 minutes compared with 3 to 6 minutes?

Space for calculation

__________ times **1**

Marks

9. African wild dogs are social animals that hunt in packs. They rely on stamina to catch grazing prey such as wildebeest.

The table shows the effect of wildebeest age on the average duration of successful chases and the percentage hunting success.

Wildebeest age	*Stage*	*Average duration of successful chases* (s)	*Hunting success* (%)
up to 1 year	calves	20	75
from 1 – 2 years	juveniles	120	50
over 2 years	adults	180	45

(*a*) Describe the effect of wildebeest age on the average duration of successful chases.

____

______________________________ **1**

(*b*) How many times longer does it take the wild dogs on average to successfully hunt adult wildebeest rather than calves?

Space for working

__________ times **1**

(*c*) Suggest a reason why hunting success is greatest with calves.

____

______________________________ **1**

(*d*) Wild dogs kill a greater number of adult wildebeest than calves.

Explain this observation in terms of the economics of foraging behaviour.

____

____

______________________________ **1**

(*e*) State an advantage of cooperative hunting to the wild dogs.

______________________________ **1**

(*f*) Following a successful hunt, wild dogs may be displaced from their kill by spotted hyenas. What type of competition does this show?

______________________________ **1**

Marks

10. (*a*) Fulmars and Common Terns are seabirds that breed in large social groups.

The table compares features of breeding in these birds.

Feature of breeding	*Fulmar*	*Common Tern*
nest distribution and location	crowded on cliff ledges	scattered on pebble beaches
egg number and colour	single white egg	three speckled eggs
chick behaviour	remains in nest until able to fly	can move short distances from nest soon after hatching

(i) Use information in the table to explain why Fulmars are less vulnerable to predation than Common Terns.

____

____

______________________________ **1**

(ii) Suggest how features of Common Tern eggs and chicks may increase their survival chances.

1 Eggs ______________________________

____

______________________________ **1**

2 Chicks ______________________________

____

______________________________ **1**

(*b*) Explain how living in large social groups may help animals in defence against predators.

____

____

______________________________ **1**

[Turn over

DO NOT WRITE IN THIS MARGIN

Marks

11. An investigation was carried out into the effect of lead ethanoate and calcium ethanoate on the activity of catalase.

Catalase is an enzyme found in yeast cells. It acts on hydrogen peroxide to produce oxygen gas.

The stages in the investigation are outlined below.

1 Three yeast suspensions were made by adding 100 mg of dried yeast to each of the following.

- 25 cm^3 of 0·1 M lead ethanoate solution
- 25 cm^3 of 0·1 M calcium ethanoate solution
- 25 cm^3 of water

2 The suspensions were stirred and left for 15 minutes.

3 Separate syringes were used to add 2 cm^3 of each yeast suspension to 10 cm^3 of hydrogen peroxide in 3 identical containers.

4 The volume of oxygen produced in each container was measured at 10 second intervals.

The results are shown in the table.

Time (s)	*Volume of oxygen produced* (cm^3)		
	yeast suspension + lead ethanoate	*yeast suspension + calcium ethanoate*	*yeast suspension + water*
0	0	0	0
10	6	32	38
20	10	62	56
30	14	74	78
40	15	88	86
50	16	90	88
60	17	90	90

(*a*) Why was it good experimental procedure to leave the yeast suspensions for 15 minutes at stage 2?

____

____ **1**

(*b*) Why was a separate syringe used for each yeast suspension at stage 3?

____

____ **1**

(*c*) Identify **one** variable, not already described, that should be kept constant.

____ **1**

11. (continued)

(*d*) The results for the yeast suspensions in 0·1 M calcium ethanoate and in water are shown in the graph.

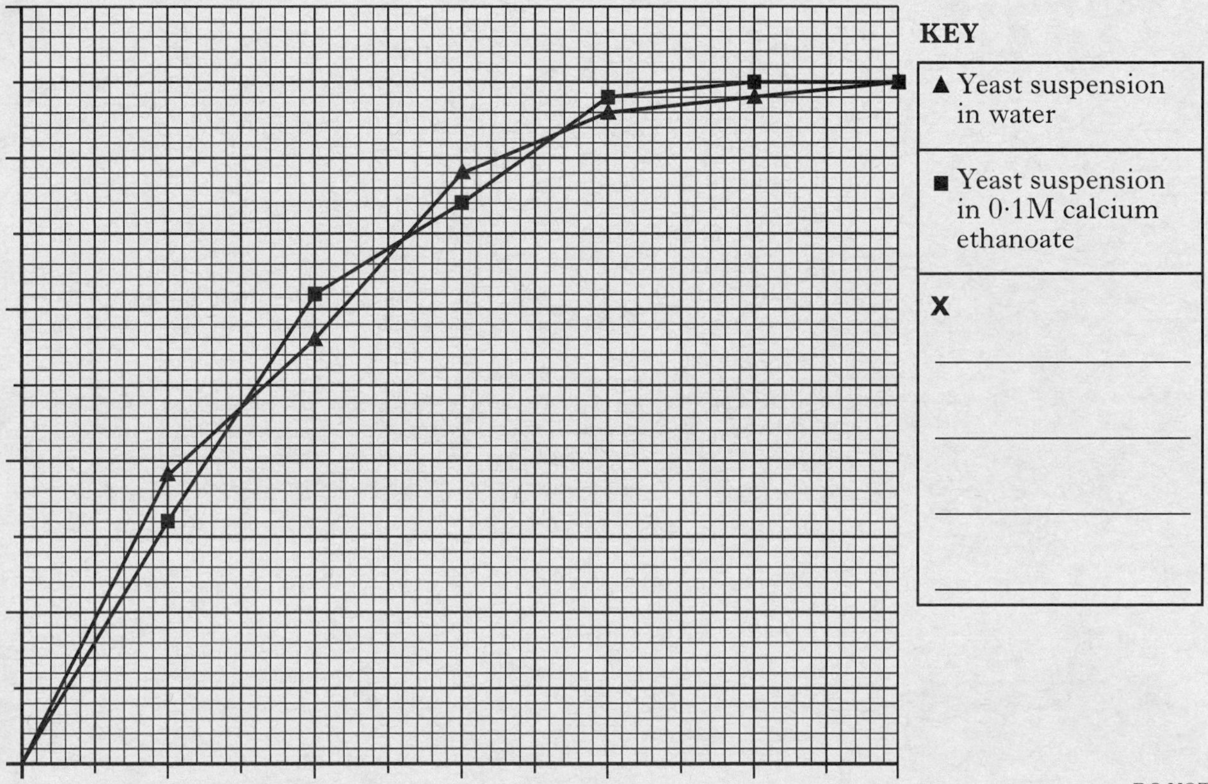

KEY

▲ Yeast suspension in water

■ Yeast suspension in 0·1M calcium ethanoate

X

DO NOT WRITE IN THIS MARGIN

Marks

Use information from the table to complete the graph by:

(i) adding the scale and label to each axis; **1**

(ii) presenting the results for the yeast suspension in 0·1 M lead ethanoate **and** completing the key. **1**

(Additional graph paper, if required, will be found on page 36.)

(*e*) Explain how the yeast suspension in water acts as a control.

____

____

____ **1**

(*f*) Draw **two** conclusions from the results in the table.

1 ____

____ **1**

2 ____

____ **1**

DO NOT WRITE IN THIS MARGIN

Marks

12. (*a*) The diagram shows a section through a barley grain.

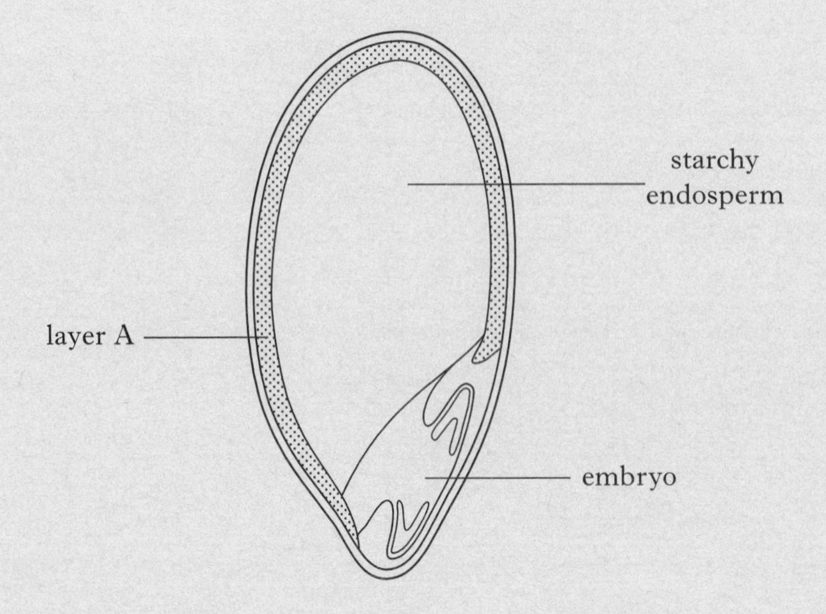

Layer A produces α-amylase.

(i) Name layer A.

______________________________ **1**

(ii) What substance made by the embryo induces α-amylase production?

______________________________ **1**

(iii) Explain the role of α-amylase in the process of germination.

____

____

______________________________ **2**

(*b*) Give **one** practical application of a plant growth substance.

______________________________ **1**

DO NOT WRITE IN THIS MARGIN

Marks

13. The graph shows the results of an investigation into the relationship between environmental temperature and body temperature for a bobcat and a rattlesnake.

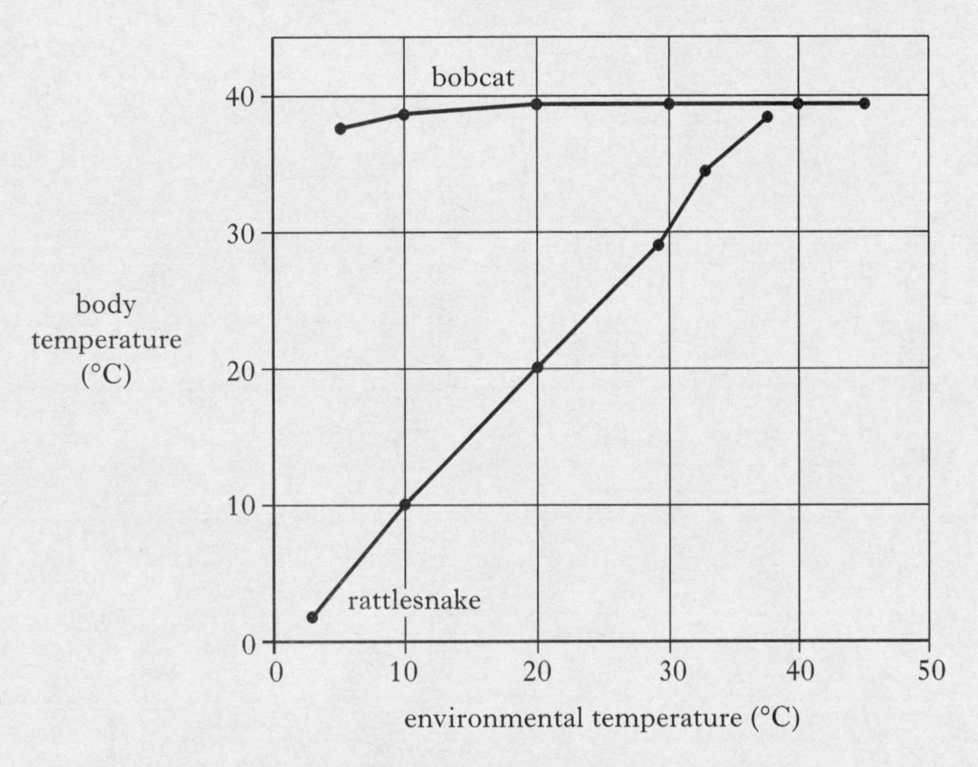

(*a*) Using information from the graph, **underline** one of the alternatives in each pair to make the sentence correct.

The rattlesnake is an {ectotherm / endotherm} because the results show that it {can / cannot} control its body temperature. **1**

(*b*) Describe a rattlesnake behaviour pattern that is likely to raise its body temperature above the surrounding air temperature.

____

____ **1**

(*c*) What evidence from the graph suggests that the bobcat has mechanisms to prevent overheating?

____

____ **1**

(*d*) Explain why the bobcat's metabolic rate is greater at 10 °C than at 30 °C.

____

____

____ **2**

SECTION C

Marks

Both questions in this section should be attempted.

Note that each question contains a choice.

Questions 1 and 2 should be attempted on the blank pages which follow.

Supplementary sheets, if required, may be obtained from the invigilator.

Labelled diagrams may be used where appropriate.

1. Answer **either** A **or** B.

A. Give an account of populations under the following headings:

(i) the importance of monitoring wild populations; **5**

(ii) the influence of density-dependent factors on population changes. **5**

(10)

OR

B. Give an account of growth and development under the following headings:

(i) the influence of pituitary hormones in humans; **4**

(ii) the effects of Indole Acetic Acid (IAA) in plants. **6**

(10)

In question 2, ONE mark is available for coherence and ONE mark is available for relevance.

2. Answer **either** A **or** B.

A. Give an account of the absorption of light energy by photosynthetic pigments and the light-dependent stage of photosynthesis. **(10)**

OR

B. Give an account of the structure of RNA and its role in protein synthesis. **(10)**

[*END OF QUESTION PAPER*]

DO NOT WRITE IN THIS MARGIN

SPACE FOR ANSWERS

DO NOT WRITE IN THIS MARGIN

SPACE FOR ANSWERS

DO NOT WRITE IN THIS MARGIN

SPACE FOR ANSWERS

SPACE FOR ANSWERS

DO NOT WRITE IN THIS MARGIN

SPACE FOR ANSWERS

SPACE FOR ANSWERS

SPACE FOR ANSWERS

DO NOT WRITE IN THIS MARGIN

SPACE FOR ANSWERS

ADDITIONAL GRAPH PAPER FOR QUESTION 11(*d*)

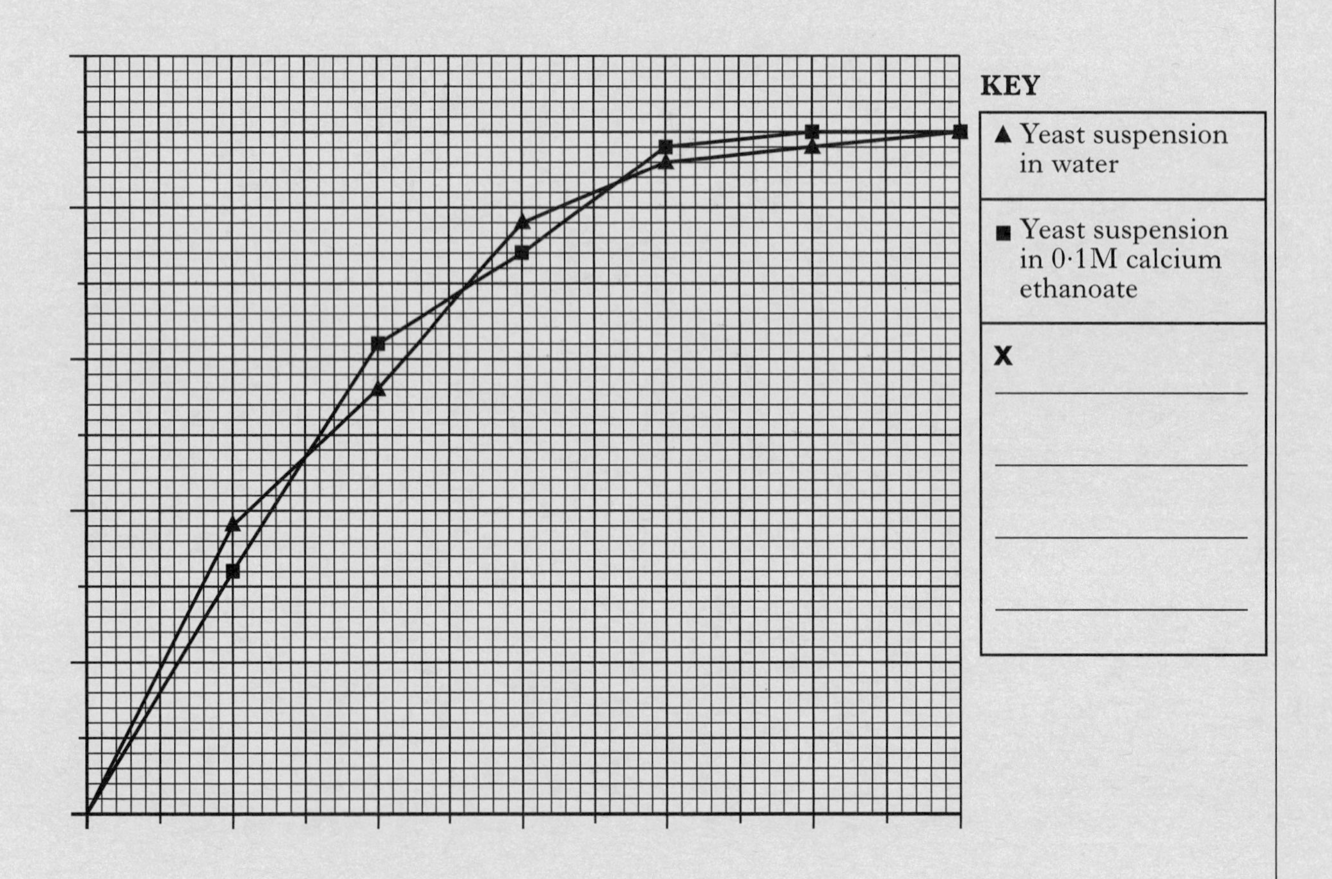

[BLANK PAGE]